Professional Collaborations in Mathematics Teaching and Learning

Seeking Success for All

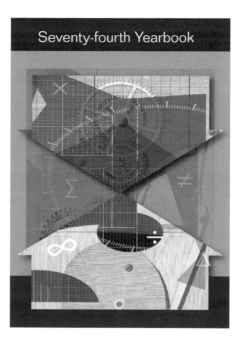

Seventy-fourth Yearbook

Jennifer M. Bay-Williams
Seventy-fourth Yearbook Editor
University of Louisville
Louisville, Kentucky

William R. Speer
General Yearbook Editor
University of Nevada, Las Vegas
Las Vegas, Nevada

NCTM® | NATIONAL COUNCIL OF
TEACHERS OF MATHEMATICS

ISSN 0077-4103
ISBN 978-0-87353-697-4

The National Council of Teachers of Mathematics is a public voice of
mathematics education, supporting teachers to ensure equitable mathematics learning
of the highest quality for all students through vision, leadership, professional
development, and research.

Printed in the United States of America

Contents

IV. Teacher Leaders and Coaches.................................... 145
How can we develop teachers into leaders within and across schools?

VII. Sustaining Professional Collaborations 283
What must be in place to sustain the energy and gains within a professional collaboration?

Preface

Collaboration with colleagues is essential in many, if not most, professions. We, as educators, are professionals and often benefit from the insights, the expertise, and the collective efforts of our colleagues. Collaboration in education is not a frill: it is an essential component of our practice.

The 2012 Yearbook of the National Council of Teachers of Mathematics (NCTM) offers an opportunity and a resource for exploring a wide variety of perspectives on the nature and design of professional collaborations as they relate to mathematics teaching and learning. This includes examining such elements as the composition of a collaborative partnership; the role of various stakeholders; and traditional and nontraditional means of promoting and influencing teacher partnerships such as coaching and mentoring.

This year's Yearbook panel sought to avoid simply gathering anecdotes, stories, and vignettes, however valuable such illustrations might be in focusing on the unique challenges to be encountered in establishing effective learning and teaching collaborations. The goal was to structure a Yearbook that would open the door to understanding the concepts underlying successful collaborations, to establish the principles that lead to success, and to search for patterns of design and structure that offer state-of-the-art thought and direction both to those making a first attempt at collaboration and to those revising an earlier effort. Rhetoric is usually less effective than concrete ideas for implementation, so this book is designed to offer a perspective to the reader that reinforces reflection leading to action. There is nothing inherently magic about collaboration and professional learning communities—it is the thought that goes in that defines the benefits that come out.

In chapter VI of *Alice's Adventures in Wonderland,* Lewis Carroll (the pen name of the mathematician Charles Dodgson) wrote the following dialogue between Alice and the Cheshire Cat:

> "Would you tell me please, which way I ought to go from here?"
>
> "That depends a good deal on where you want to go," said the Cat.
>
> "I don't care much where—" said Alice.
>
> "Then it doesn't matter which way you go," said the Cat.

We do much care where we go, and it does matter which way we go. This Yearbook is designed to assist you in sorting out the various paths to success and in charting your course toward establishing, adjusting, redesigning, or reinventing professional learning communities. As you read this book, choose a pathway that leads you to the solutions you require for the following questions:

- What does your situation require in terms of professional collaborations?
- What would it take to implement just one idea from these chapters?
- How can you build support for creating a learning community?
- Which of the strategies for collaboration makes the greatest sense in light of your goals and context?
- What approaches to collaboration appeal to you the most?
- How will you measure the effectiveness of your collaborative actions?

The Seventy-fourth Yearbook of the NCTM, *Professional Collaborations in Mathematics Teaching and Learning: Seeking Success for All*, consists of twenty-two chapters, divided into seven parts. The three chapters in Part I present an overview of the conceptual and theoretical underpinnings of professional collaborative frameworks. The ten chapters in the following three parts ("Students and Student Learning," "Teachers and Teacher Learning," and "Teacher Leaders and Coaches") explore the dimensions and interactions of learning associated with collaborations. The final three parts of the book contain nine chapters that delve into pragmatic suggestions for promoting and assessing long-term change and valuing the role of collaborations in daily teaching. Together, these twenty-two chapters offer sound research-driven solutions to address the challenges that we face as educators, as well as extensive tools and practical strategies for the implementation and evaluation of professional collaborations. In addition, you'll find suggested resources and descriptions of multiple experiences that can be used to help refine collaborations and support their sustainability.

This yearbook provides us with a vision of what professional collaborations have been, are currently, and can become. Lewis Carroll also wrote the following dialogue between the White Rabbit and the King:

"Where shall I begin, please your Majesty?" he asked.

"Begin at the beginning," the King said gravely, "and go on till you come to the end: then stop."

In our search for the most perfect professional collaboration, we most likely will never believe that we have reached the end the King speaks of and then stop. If you are at the beginning, or in the middle of, a collaboration to improve student or teacher learning, then the contents of this yearbook are intended to help you in that journey.

......

Producing an NCTM yearbook would be an insurmountable task without the help and guidance of many people. One of the most challenging tasks in putting together a yearbook is to select from the multitude of manuscripts submitted in response to the original call. As is the custom, the selection and organization of these manuscripts eventually falls heavily on the general editor and the volume editor, but the Editorial Panel represents the heart and soul of the process. I am indebted to the Editorial Panel for the 2012 Yearbook for their deep expertise and a wide body of experiences across the spectrum of professional learning communities. My special thanks go out to the following panel members for their consistent demonstration of insight, knowledge, creativity, and command of the tasks before them.

The 2012 NCTM Yearbook Editorial Panel

Jennifer M. Bay-Williams

Jenny Bay-Williams, the 2012 Yearbook Editor, is a professor of mathematics education at the University of Louisville (Kentucky). She previously taught elementary, middle, and high school in Missouri and in Peru. Bay-Williams's contributions to NCTM include numerous articles in the journals *Teaching Children Mathematics*, *Mathematics Teaching in the Middle School* (where she also edited three different departments), and *Mathematics Teacher*, as well as several books (*Developing Essential Understanding of Addition and Subtraction for Teaching Mathematics in Prekindergarten–Grade 2; Navigating through Mathematical Connections in Grades 6–8;* and *Growing Professionally*). Beyond NCTM, she is the co-author of *Elementary and Middle School Mathematics: Teaching Developmentally* and its related *Field Experience Guide; Math and Literature: Grades 6-8;* and *Math and Nonfiction: Grades 6-8*. Bay-Williams is a past president of the Association of Mathematics Teacher Educators (AMTE) and serves on the Board of Directors for TODOS: Mathematics for All.

Antonia Cameron

Toni Cameron is a mathematics educator and writer living in New York City. As a mathematics coach and professional developer, she currently works nationally and internationally with teachers, principals, and coaches. As a writer, Cameron has co-authored fifteen facilitator guides in the professional development series *Young Mathematicians at Work*; she has also written several of the Contexts for Learning units published by Heinemann. As co-director of the Mathematics in the City Project (MitC), City College of the City University of New York, she worked locally and internationally to assist educators through college courses, summer institutes, keynotes, workshops, and on-site staff development. Cameron is currently partnering with Lucy West at Metamorphosis Teaching Learning Communities (www.metatlcinc.com), where their work focuses on how to use collaborative learning communities and coaching as vehicles for creating systemic change in schools and districts. They are currently working on a book about high-leverage content coaching and teacher leadership (Heinemann, in press).

Florence Glanfield

Florence Glanfield is an associate professor of mathematics education in the Department of Secondary Education at the University of Alberta in Edmonton, Alberta, Canada. Glanfield has worked with students and teachers in all geographic regions of Canada, developed provincial mathematics curriculum, participated in the implementation of curriculum, and developed student assessment materials. She is actively involved in provincial, national, and international mathematics education organizations, and she has authored or co-authored numerous teacher and student resources and research articles. Committed to collaboration in research, she is currently engaged in five international, national, and local collaborative research projects. Glanfield and two colleagues together received the 2010 Alberta Teacher's Association Educational Research Award for their work in a collaborative research project in mathematics education with an indigenous community.

Richard S. Kitchen

Rick Kitchen is a professor of leadership in mathematics education in the College of Education and the Department of Mathematics and Statistics at the University of New Mexico (UNM). His primary research interests are in equity and diversity in mathematics education, as well as in formative assessment formats for English language learners in mathematics. Kitchen is the lead author of *Mathematics Education at Highly Effective Schools That Serve the Poor: Strategies for Change* (Lawrence Erlbaum Associates, 2006). He was also the co-Principal Investigator at UNM for the National Science Foundation–funded Center for the Mathematics Education of Latinos/as (CEMELA). Kitchen is the founder and director of Escuela Luz del Mundo, a college-prep Christian school that serves the poor.

David K. Pugalee

David Pugalee is a professor of education at the University of North Carolina at Charlotte, where he currently serves as director of the Center for Science, Technology, Engineering, and Mathematics Education. Pugalee was an author of several of the middle grades Navigations books published by NCTM. His work primarily focuses on language and communication in mathematics and includes numerous journal articles and books on the topic. He recently conducted NCTM e-workshops on Response to Intervention in mathematics, and he strives in his work to promote quality STEM education pre-K–16.

Amy Roth McDuffie

Amy Roth McDuffie is a mathematics educator and researcher in the College of Education at Washington State University–Tri-Cities.

Roth McDuffie researches preservice and practicing teachers' professional learning and development, with a focus on uses of curriculum and student equity. She has published articles in *Teaching Children Mathematics*, *Mathematics Teaching in the Middle School*, *Mathematics Teacher*, *Journal of Mathematics Teacher Education*, and *School Science and Mathematics*, as well as other articles and book chapters in mathematics education and teacher education publications. Roth McDuffie enjoys collaborating with teachers to support and study those teaching practices that are aimed at improving students' learning.

In particular, this yearbook would not have been possible without the outstanding contributions of the volume editor, Jennifer Bay-Williams, of the University of Louisville. Jenny has been a keystone for all of us on this project. Her extreme dedication and sense of duty and commitment have helped us all in ways that are most visible in the quality of the final product.

I would be remiss if I failed to acknowledge the stellar support and guidance received from NCTM. Several individuals worked behind the scenes to ensure the prompt and quality production of the volume. These include Joanne Hodges, Director of Publications; Myrna Jacobs, Publications Manager; Larry Shea, Copy and Production Editor; Randy White, Production Manager; Ken Krehbiel, Associate Executive Director for Communications; Kathe Richardson, Meeting Planner; and Linda Cooper Foreman, Educational Materials Committee Chairperson during the production of this Yearbook. These dedicated people were there every step of the way and provided much direction and support on process and procedure. In fact, many other unnamed members of the NCTM staff worked long and hard to bring this project to fruition. My sincere thanks go out to them all for their help and direction in making this yearbook a reality.

Dr. William R. Speer
NCTM General Editor, 2011–2013 Yearbooks

Part I
Frameworks for Professional Collaborations

Which components of collaborations matter the most?

Educators cannot make it through a school day without collaborating with others. We work together to resolve myriad issues related to everything from student learning to schedules. Committees are formed to address ongoing or pervasive issues. Teachers and university educators are engaged in constant professional development, whether it is reading an article, attending a workshop or conference, or observing a colleague. We all head into our interactions seeking to be better at our craft, and better at positively impacting the students and teachers that we encounter.

Collaboration, learning, and professional growth are all worthy practices, yet they do not necessarily constitute a professional learning community or a professional collaboration focused on learning. It is important to consider what elements of a collaboration make it a *learning* community. In other words, what components must be prevalent in order to have an effective collaboration focused on learning?

As you read the chapters in this section, written by educators deeply engaged in defining, refining, and advocating for professional learning communities, use the following questions to consider the essential question about the frameworks used for collaborations: Which components really matter? These questions can be fodder for your professional learning community or for your own individual reflection.

- At what point does a collaboration or group actually become a professional learning community?
- Which of the components discussed is poorly understood? Difficult to establish? A good starting point?
- In what way can the cases shared in these chapters help me/us to improve the function of our collaboration?

CHAPTER 1

Improving Mathematics Achievement

The Power of Professional Learning Communities

Robert Eaker
Janel Keating

F OR WELL OVER a half a century, educators and politicians in the United States have recognized the need for improved student achievement in mathematics. This need has been a topic of public and political concern, and both teachers and students have experienced initiative after initiative, and reform after reform, in attempts to fulfill it. Despite these repeated efforts, however, the performance of students in mathematics remains an area of considerable concern and attention.

The Need for Professional Learning Communities

Generally speaking, efforts in the United States to improve student performance in mathematics fit into three categories. The most frequent approach, and the one with the most widespread implementation, has been to require more mathematics for both students and teachers. Many states have changed high school graduation requirements to include more mathematics, and the licensure requirements for teachers have been altered to require more content courses in mathematics.

Additionally, periodic calls have been made not only to require that students learn more mathematics, but also to require that the conceptual foundations of mathematics be taught much earlier. For example, the Common Core State Standards (CCSS), which have been adopted in a large majority of states, include rational number concepts at earlier grades than was previously the case. Many mathematics textbooks, electronic resources, and assessment products have been developed or adapted to reflect the changes recommended in standards documents.

Other recommendations and initiatives have aimed at improving the instructional practices of teachers who teach mathematics from elementary school through high school. For example, the *Principles and Standards for School Mathematics* (NCTM 2000) outline effective instructional strategies. In addition, professional development efforts, including college courses, have emphasized new and more effective instructional strategies for the teaching of mathematics.

It would seem that all bases—curriculum, instruction, and assessment—have been addressed. After all, new and more rigorous standards have been adopted. State curricular standards have been modified to ensure that students are exposed to important mathematical concepts, and teachers have been provided with training and resources to enable them to become more effective in their teaching of mathematics. Why is it, in spite of decades of requiring *more* mathematics for both students and teachers, through *more* rigorous curricula, taught at an *earlier* age, by teachers who utilize *improved* instructional strategies, achievement levels of American students still remain lacking?

Structure Alone Is Not Enough: The Power of Organizational Culture

One reason we have failed to reap the benefits of repeated efforts to improve mathematics achievement is that most efforts have been aimed at changing state, district, and school *structures*—more rigorous standards, more rigorous graduation requirements, more rigorous licensure requirements, more relevant and rigorous curricula, and improved professional training and development. The emphasis on changing school policy structures is somewhat understandable. Structural changes, such as strengthening high school graduation requirements or teacher licensure requirements, can be implemented rather quickly, and "raising standards" is always politically popular. Unfortunately, *structural changes alone are rarely enough.* As Seymour Sarason writes, "To put it as succinctly as possible, if you want to change and improve the climate and outcomes of schooling for both students and teachers, there are features of the school culture that have to be changed, and if they are not changed, your well-intentioned efforts will be defeated" (1996, p. 340).

Perhaps it is time to look through a different lens, the lens of the school and district culture in which mathematics teachers teach, and the culture in which students learn. While paying attention to both structure and culture is necessary, inadequate attention has been given to the professional "work-life" of teachers—the context and culture in which they work. Research supports this observation. For example, Roland Barth (2001) concludes, "The school's culture dictates, in no uncertain terms, 'the way we do things around here.' Ultimately, a school's culture has far more influence on life and learning in the schoolhouse than the state department of education, the superintendent, or even the principal can ever have . . . The culture is the historically transmitted pattern of meaning that wields astonishing power in shaping what people think and how they act" (pp. 7–8).

Richard DuFour, Rebecca DuFour, and Robert Eaker (2008) define *culture* as the assumptions, beliefs, values, expectations, and habits that constitute the norm for an organization. We observe that all schools have cultures. However, certainly not all school cultures are alike or equally effective. We write, "They [cultures] may foster collaboration or isolation, promote self-efficacy or fatalism, be student centered or teacher centered, regard teaching as a craft that can be developed or as an innate art, assign primary responsibility to teachers or students, view administrators and teachers as colleagues or adversaries, encourage continuous improvement or defense of the status quo, and so on" (p. 90).

The observation that attention must be paid to district and school culture leads to an obvious question—what kind of culture would best impact and support the difficult and complex work of teachers? More specifically, what kind of culture would best impact mathematics teaching and learning? We believe the answer to this question has never been clearer. *Our best hope for sustained school improvement, and the improvement of mathematics achievement, lies in district and school cultures that reflect the concepts and practices of schools that function as professional learning communities.*

Broad Support for Professional Learning Communities

Rarely in the history of American public education has there been such widespread agreement regarding the best path to improved student learning. Consider this small sample of researchers' calls for professional learning communities:

"If schools want to enhance their organizational capacity to boost student learning, they should work on building a professional learning community that is characterized by shared purpose, collaborative activity, and collective responsibility among staff" (Newmann and Wehlage 1995, p. 37).

"[We recommend that] schools be restructured to become genuine learning organizations for both students and teachers; organizations that respect learning, honor teaching, and teach for understanding" (Darling-Hammond 1996, p. 198).

"The framework of a professional learning community is inextricably linked to the effective integration of standards, assessment, and accountability. . . . The leaders of professional learning communities balance the desire for professional autonomy with the fundamental principles and values that drive collaboration and mutual accountability" (Reeves 2005, pp. 47–48).

"Well-implemented professional learning communities are a powerful means of seamlessly blending teaching and professional learning in ways that produce complex, intelligent behavior in all teachers" (Sparks 2005, p. 156).

"A school-based professional learning community can offer support and motivation to teachers as they work to overcome tight resources, isolation, time constraints and other obstacles they commonly encounter. . . . In schools where professional community is strong, teachers work together more effectively, and put more effort into creating and sustaining opportunities for student learning" (Kruse, Seashore Louis, and Bryk 1994, p. 4).

"Such a tipping point—from reform to true collaboration—could represent the most dramatic shift in the history of educational practice. . . . We will know we have succeeded when the absence of a 'strong professional learning community' in a school is an embarrassment" (Schmoker 2004, p. 431).

"We have an increasingly clear picture of the nature and importance of professional learning communities in schools. We now understand that such communities do not merely represent congeniality. Rather, they dig deeply into learning. They engage in disciplined inquiry and continuous improvement in order to 'raise the bar' and 'close the gap' of student learning and achievement" (Fullan 2005a, p. 209).

Principles and Standards for School Mathematics (NCTM 2000), in a discussion of how to implement the Teaching Principle, presents a strong argument for the need for learning communities:

> Most mathematics teachers work in relative isolation, with little support for innovation and few incentives to improve their practice. Yet much of teachers' best learning occurs when they examine their teaching practices with colleagues. . . . Too often we place the responsibility for change solely on the shoulders of teachers and then blame them when things do not work as expected. We need instead to address issues in a systemic way, providing teachers with the resources they need for professional growth. . . . The typical structures of teachers' workdays often inhibit community building, but structures can be changed. In some cultures, shared discussions of students and teaching are the norm. . . . During "lesson study," teachers plan the lesson, teach the lesson with colleagues watching, revise the lesson collaboratively, teach the revised lesson, evaluate and reflect again, and share the results in written form. . . . Although this level of professional collaboration may be hard for U.S. and Canadian teachers to imagine within the constraints of the prevailing professional culture and system, it illustrates the potential power of learning communities to improve mathematics teaching and learning. Finding ways to establish such communities should be a primary goal for schools and districts that are serious about improving mathematics education. (pp. 370–71)

The National Council of Supervisors of Mathematics (NCSM) has also emphasized the need for schools to function as professional learning communities. Their standards for ensuring equity in mathematics achievement are based on the benchmark that "every teacher works interdependently in a collaborative learning community to erase inequities in student learning" (NCSM 2008, p. 18). In order to meet this standard, NCSM calls for leaders who will—

- recognize, understand, and model the components of a professional learning community;

- develop a results-driven culture, and participate in a collaborative team for mathematics program improvement;

- collaborate with teachers to create the support and structures necessary to implement a professional learning community;

- support the learning of teachers to work as a professional learning community in order to monitor gains in student achievement for every student population; and

- advocate for and ensure a systemic implementation of a professional learning community throughout all aspects of the mathematics curriculum, instruction, and assessment at the school, district, regional, or provincial level.

Beyond mathematics education, many other educational organizations, such as the National Science Teachers Association (NSTA), have endorsed the professional learning community concept.

Imagine Teaching Mathematics in This School

What would it be like to teach in a school that functions as a high-performing professional learning community? Perhaps the most striking difference would be this: rather than teaching in a culture of teacher isolation, all teachers in the school would be expected to be contributing members of collaborative teams. In a school that functions as a professional learning community, great care is given to the makeup of each team. The primary organizing idea is that team members should generally teach the same content so that discussions can focus on what students are learning and how instructional strategies are (or are not) affecting that learning. For example, an elementary teacher may be on the third-grade team and/or on a pre-K–second-grade vertical mathematics team. At the middle school or high school level, a teacher may be a member of the math team, or in larger schools, she may be a member of the math team, but she will spend the majority of her time with her course-specific team—for example, an algebra team or a geometry team. The point is, teachers would have a shared purpose for being on a collaborative team, and that purpose would be *to improve student learning*! In the sections that follow, we share the elements that are essential for a successful professional learning community and weave them into the story of White River School District in Buckley, Washington.

The Work of a Team

What collaborative team members actually do in a professional learning community goes far beyond simply having conversations. The team engages in purposeful work toward improving teaching and learning. Early on, each team develops specific organizational and structural foundations for their future work. For example, they will collaboratively develop *team norms*—guidelines for how the team will do their work day in and day out. In White River, as in all successful professional learning communities, it was understood that unless each team collaboratively set parameters to guide their work, discussions would soon drift from the substantive issues related to improving student learning. Each team collaboratively developed team norms or protocols such as: "We will start our meetings on time and end on time," "We will always work from an agenda, and we will stick to agenda items," and "We will value each other's comments by practicing good listening skills." Teams in White River School District frequently refer to their norms, using them to improve the quality of their interactions.

Determining Curricular Goals

The teams also analyze state and national standards for mathematics, identifying what must be learned, as well as how it might be taught. In schools that function as professional learning communities, collaborative teams of teachers literally become *students of the standards*, clarifying the meaning of each standard and developing common pacing guides to ensure adequate time is being given to the "essential" learning outcomes—or as Doug Reeves (2002) refers to them, the "power standards."

Through the work of collaborative teams, mathematics teachers develop what Marzano (2003) refers to as a "guaranteed and viable" curriculum. Simply put, students must be *guaranteed* the same mathematics curriculum regardless of the teachers they are assigned or the school they attend. Additionally, students would be taught a *viable* curriculum because teacher teams would have collaborated to determine the amount of time needed to effectively teach each standard.

White River School District is an example of a school district that has successfully embedded the work of a professional learning community into each school. Each team ensures a "guaranteed and viable" curriculum for each student regardless of the school they attend or the teacher to whom they are assigned. For example, the geometry team at White River High School, led by Cody Mothershed, regularly engages in discussions about the essential outcomes that every student should know or be able to do, continually sharpening the focus of their teaching. Figure 1.1 illustrates the product of their collaborative work for Unit One.

Power Standards

G.1.F: Explain the role of definitions, undefined terms, postulates (axioms), and theorems. We broke this standard into two: G.1.F (definitions) and G.1.F (postulates)

G.2.B: Know, prove, and apply theorems about angles.

G.2.D: Describe the intersection of lines in the plane and in space, of lines and planes, and of planes in space.

G.4.B: Determine the coordinates of a point that is described geometrically.

Learning Targets

- Students will recognize and use correct notation for lines, rays, segments, planes, points, and angles.
- Students can describe the intersection of two lines, two planes, and a plane and a line not in the plane and give an example of each using objects in the classroom.

Fig. 1.1. White River High School geometry team plans for Unit One

> ### Learning Targets—Continued
>
> - Compare notation for angles and segments for when the measurement is talked about compared to the segment or angle itself.
> - Describe the Segment Addition Postulate using words and examples and apply the postulate to solve mathematical problems.
> - Students can compare the Angle Addition Postulate to the Segment Addition Postulate and be able to use both to solve mathematical problems.
> - Recognize and know key properties of linear pair, adjacent, vertical, complementary, and supplementary angles.
> - Students can describe midpoint and find midpoint using a couple different methods.

Fig. 1.1—*Continued*

Determining Learning Outcomes

Once teams, and ultimately the entire White River School District, develop a "guaranteed and viable" mathematics curriculum, the teams begin to drill deeper into student learning outcomes. They focus on questions such as, "What would student work look like if this particular standard was met?" Even when teachers align their instruction to the standards, if there is no collaboration around the implementation of the standards there can be a wide disparity between teachers' expectations for high-quality student performance. Developing rubrics for high-quality student work naturally leads to the discussion of the common scoring of student products, and ultimately to discussions about the reporting of pupil progress.

At White River High School, once the geometry team had collaboratively agreed upon the power standards for Unit One, they were then able to focus their discussions on such topics as pacing, agreeing on what student work would look like if the standards were met, appropriate homework assignments, and common scoring and assessment formats and conditions.

Importantly, district leaders in White River directed the work of the collaborative teams. For example, teams did not have the latitude to choose whether or not they should identify and add meaning to state power standards, or collaboratively develop and analyze the results of common formative assessments, or develop specific ways to provide students with additional time, support, and enrichment within the school day. On the other hand, teams had great latitude in *how* they approached these tasks.

Of course, disagreements often occur when engaging in a process to identify state "power standards." After all, it is a collaborative process. When disagreements occur in White River, team members refer to the state summative assessments, asking, "What are the 'big things' students will be asked to do in order to demonstrate proficiency?" Additionally, the curriculum guides in White River are constantly being reviewed and tweaked.

Developing and Using the Results of Common Formative Assessments

As noted earlier, collaborative teams are not content with merely clarifying what students must learn. The teams focus on the question, "How will we know if our students are learning?" Collaborative teams develop common formative assessments in order to monitor the learning of each student, skill by skill, on a frequent and timely basis. They pursue a cultural shift from an overreliance on summative assessments toward frequent and collaboratively developed common formative assessments. Simply put, being a mathematics teacher in a professional learning community means being part of a collaborative team that recognizes students are more apt to perform well on high-stakes summative assessments *if the quality of their learning is regularly monitored along the way*—especially when the results of the formative assessments are used to provide students with additional time, support, or enrichment.

It is not so much that team members in professional learning communities develop common formative assessments; the power of formative assessments lies in how they are utilized. Teachers collaboratively analyze student learning data, as well as examples of student work. They discuss assessment results, item by item, as well as the effectiveness of the assessment itself. Teachers highlight strengths in student learning and identify areas of concern. They monitor the learning of each student, skill by skill. Most important, they collaboratively decide on appropriate interventions for students and future plans for instruction.

As a result of their collaborative analysis of the results of their formative assessments and an examination of student work, the geometry team at White River High School identified specific areas in which students struggled, as well as an analysis of the assessment, as described in figure 1.2.

Student Difficulties

- Describing the intersection of the lines and planes
- Using the correct notation for all the geometric terms
- Solving for the other endpoint of a segment given one endpoint and the midpoint
- Incorrectly setting up an equation by getting congruency mixed up with the addition postulates

Fig. 1.2. White River High School geometry team identifies student learning needs

> **Analysis of the Assessments**
>
> - The wording of the questions may have confused students. Further evidence must be gathered in order to determine if students were confused by the wording, or if they just did not know the concepts.
> - Students seem to have difficulty with solving equations (algebraic concepts), which led to an inability to solve the geometry problems.
> - This unit introduced many new terms and vocabulary. Students are using the wrong terms in many cases, leading to errors. We need to emphasize these terms and phrases throughout the year to help students remember and appropriately apply the vocabulary.

Fig. 1.2—*Continued*

Additional Time and Support

Students learn at different rates and in different ways, so a school that functions as a professional learning community will recognize that some students will need additional time and support, as well as encouragement. In schools where teaching is an isolated practice, mathematics teachers often have to figure out on their own what to do with their students who are not "getting it." Even the best-intentioned and most talented teachers can only do so much by themselves. The number of students who need additional time and support is often too great and the range of needs too wide. In schools that function as professional learning communities, systematic, schoolwide plans of "layered interventions" can be created. If a school does not provide additional time and support when students are struggling to learn, the school will fail to be effective, even if it is implementing many other effective practices.

Further, schools must recognize it is their *moral* obligation to provide support to struggling students. As DuFour, DuFour, Eaker, and Many (2010) observe, "It is disingenuous for any school to claim its purpose is to help all students learn at high levels and then fail to create a system of intervention to give struggling learners additional time and support for learning" (p. 104).

Intervention should be based on a systematic plan that is *timely* and *directive*, rather than merely "invitational." The plan should include a series of sequential "layers," in order to provide increasing support if needed. Initially, most interventions occur within the classroom. For example, a teacher simply might reteach a particular skill set, or engage students in peer tutoring. However, some students may need the benefit of more intense interventions, such as a tutor, time in a mathematics lab, or a specific supplemental mathematics program (DuFour, DuFour, Eaker, and Many 2010). The White River High School geometry team has collaboratively developed a plan to help students when they experience difficulty, as illustrated in figure 1.3.

Extra Help and Intervention

- Throughout the unit, we will use the data from the quick checks for understanding to guide our intervention and determine which students need extra help.
- The initial help for students will be provided in individual classrooms.
- Before-and-after school help (scheduled time for individual teachers to be available for students before and after school)
- STAT (Student-Teacher Access Time) [STAT time at White River High School is a specific period within the school day in which all students receive additional time, support, or enrichment.]
- Group work in which students will go through corrections and example problems
- After-unit activities to review main points of power standards
- Online textbook resources such as tutorials and practice quizzes
- Spiral review throughout year and before each comprehensive test, including:
  — Connecting current concepts with previous concepts and reviewing previous concepts while making connections
  — Using student whiteboards
  — Creating a word wall and math dictionary for vocabulary

Fig. 1.3. White River High School geometry team intervention plans

Homework and Grading

In a school that functions as a professional learning community, teaching mathematics involves discussions about topics that have a huge impact on student learning and yet are often taken as a given, such as grading and homework. Teachers make an effort to learn about effective grading and homework policies and practices. They work to gain "shared knowledge" by referring to resources, such as John Hattie's (2009) review and synthesis of hundreds of research studies on factors related to student achievement. They address such questions as, "What is the purpose of homework?" "How much homework is appropriate?" "How much weight will homework be given in grading?" and "What happens when students do not complete their homework or it is completed incorrectly?" *Teacher teams* tackle these important and complex issues that have a huge impact on student learning.

Whatever the homework issue, teams are encouraged to approach them in generally the same way. In short, a professional learning community is a "way of thinking," and this way of thinking first begins by a smaller group gaining shared knowledge by seeking out and studying "best practices."

Teacher Learning

How do mathematics teachers improve their own knowledge and skills, as well as those of their team? Schools that function as professional learning communities embrace the assumption that improved *student* learning is directly linked to improved *adult* learning, and adult learning is central to a high-performing collaborative team. As Pfeffer and Sutton (2000) observe, "The answer to the knowing-doing problem is deceptively simple: embed more of the process of acquiring new knowledge in the actual doing of the task and less in the formal training programs that are frequently ineffective. If you do it, then you will know it" (p. 27). Thus, those who teach mathematics in a professional learning community engage in professional growth experiences that arise from the learning needs of their students and their team. Their learning is truly "job-embedded."

At White River High School, decisions about professional development experiences for teachers are based on thoughtful analysis of student learning data. For example, student data revealed that students were struggling with solving one-step and two-step equations, so the algebra team realized they needed an alternate instructional method that was both tactile and concrete. District leaders provided the team with professional development in the use of algebra tiles. The team also learned how to use algebra tiles in other areas in which many of their students struggled, such as combining like terms, factoring polynomials, and adding and subtracting positive and negative numbers. District leaders in White River adopted the message of DuFour, DuFour, and Eaker (2008), who observe:

> The message is consistent and clear. The best professional development occurs in a social and collaborative setting rather than in isolation, is ongoing and sustained rather than infrequent and transitory, is job-embedded rather than external, occurs in the context of the real work of the school and classroom rather than in off-site workshops and courses, focuses on results (that is, evidence of improved student learning) rather than activities or perceptions, and is systematically aligned with school and district goals rather than random. In short, the best professional development takes place in professional learning communities. (p. 370)

The work of the geometry team at White River High School clearly illustrates this type of job-embedded learning. They worked together to identify what students should learn and created common assessments. They analyzed state test item specifications. They worked to craft questions where the cognitive complexity of the individual questions matched what students would face on the state assessment. They were clear on what standards were most important in the unit of study they would be assessing. They all agreed to the time frame when they would administer the formative assessment.

But here was the glitch: the geometry team discovered two things that greatly impacted how they graded student work and eventually how they reported pupil progress. First, teachers were scoring student work differently. For example, there were differences in scoring in such areas as how much detail students gave in proofs, or students leaving answers in improper fractions and not as mixed numbers. Teachers also graded differently when students got the correct answer but used the wrong notation. Clearly, the team needed to collaboratively develop common scoring rubrics.

Second, the team realized that some teachers spent an entire class period reviewing for the assessment, while others spent little or no time reviewing. Some teachers allowed students to use their notes, while others did not. Ultimately, the geometry team developed common scoring rubrics and team norms for preparing and administering both formative and summative assessments. These discoveries, discussions, and related changes improved student performance, as well as the performance of the teachers in their assessment practices.

Recognition and Celebration

Schools that function as professional learning communities recognize that effectively ensuring that all students learn is a difficult, complex, and incremental endeavor. The improvement process can only be sustained if those who are doing the work—day in and day out—know their hard work and incremental success are being recognized and appreciated. As Kouzes and Posner (2006) note, "There are few if any needs more basic than to be noticed, recognized, and appreciated for our efforts. . . . Extraordinary achievements never bloom in barren and unappreciative settings" (p. 44). Leaders of professional learning communities help shape the culture of schools, and ultimately the behavior of those within them, through planned, purposeful—and most importantly, *sincere and meaningful* recognition and celebration of the behaviors they value the most.

Building Effective Professional Learning Communities

Because of the success of districts such as the White River district, the idea of schools functioning as professional learning communities has become widespread across North America. In fact, the term has become so much in vogue that often any loose coupling of collaboratively developed improvement initiatives is referred to as a "learning community." As Michael Fullan (2005) has stated, "Terms travel easily … but the meaning of the underlying concept does not" (p. 67).

Beyond being a group of people, a professional learning community is *a way of thinking*—whether it is at the district, school, department/team, or individual classroom level (and preferably at all of these levels). The professional learning community concept provides a format—a framework—for connecting "best practices" into a rational, logical, and commonsense approach to achieving high levels of learning for all students and adults alike. In the sections that follow, we briefly describe elements of the framework developed by DuFour and Eaker (1998), which is evident in sites where professional learning communities have succeeded in improving student learning for all.

A Focus on Learning

The organizing idea of a professional learning community—the concept that unifies and guides everything—is the passionate belief that the fundamental purpose of schools is to move from merely ensuring that the proper content is *taught* and taught well, to ensuring that all students *learn*, and learn at high levels. When a school adopts "learning for all students" as its core mission—*and, if the administration, faculty, and staff really mean it*—virtually every aspect of the school is affected, both structurally and culturally. Andy Hargreaves (2004) describes the impact of this seismic shift by observing, "A professional learning community is an ethos that infuses every single aspect of a school's operation. When a school becomes a professional learning community, everything looks different than before" (p. 48).

DuFour and colleagues (2010) point out that functioning as a professional learning community changes *everyone*. They write, "Every educator—every teacher, counselor, principal, staff member, and superintendent—will be called upon to redefine his or her role and responsibilities. People working in isolation will be asked to work collaboratively. People accustomed to hoarding authority will be asked to share it. People who have operated under certain assumptions their entire careers will be asked to change them" (p. 248). This is accomplished by focusing *intensely* on four fundamental questions.

1. *If* we believe the fundamental purpose of schools is learning, just what is it we expect all students to learn? Collaborative teams of teachers *clarify* and *prioritize* the essential outcomes in every subject, grade, and course.

2. *If* we believe the fundamental purpose of schools is learning, and we are clear about what it is we expect students to learn, how will we know if they have learned it? Collaborative teams of teachers analyze student learning—student by student, skill by skill. They do this primarily through the use of collaboratively developed common formative assessments.

3. *If* we believe the fundamental purpose of schools is learning, and we are clear about what it is we expect them to learn, and we have a system in place to monitor the learning of each student, how will we, as a school, respond when students experience difficulty with their learning? In a professional learning community, a systematic plan ensures students receive additional time and support within the school day, regardless of their teacher.

4. *If* we believe the fundamental purpose of schools is learning, and we are clear about what we expect students to learn, and we have a system in place to monitor the learning of each student, how will the school respond when students *do* learn the essential outcomes? Professional learning communities seek to "stretch" the aspirations and performance levels of *all* students. Students who demonstrate proficiency are stretched by a systematic plan for enrichment. Additionally, when students experience improvement, both the students and their teachers are recognized and celebrated for their accomplishments.

All the other characteristics of a professional learning community that follow below flow from this fundamental shift.

High-performing, collaborative teams

Schools that function as professional learning communities go beyond merely organizing the faculty and staff into collaborative teams. There is an intense focus on what the teams actually *do*. For example, teams are expected to clarify essential learning outcomes, collaboratively analyze student learning data and products (particularly the results of formative common assessments) and reflect on the effectiveness of their instructional practices. In short, teams accept *collective responsibility for results*, and are continually seeking to improve their own performance as well as the performance of their students.

Collective inquiry: Seeking "best practice"

In professional learning communities, people realize that when teachers "average opinions" it simply means they agreed. It does not mean they have found the *best* solution or approach to a problem. Teams in a professional learning community approach a problem or issue by first "seeking shared knowledge"—studying the "best that is known" about the particular topic being addressed. Just as we expect that what a physician does be based on the *best* knowledge available at any particular time, it is an educator's responsibility to learn new knowledge and to exhibit behaviors that are congruent with the most current knowledge base.

How do teams seek out and find "best practice"? One way is within the collaborative team itself. If teams of teachers are analyzing student learning, and find that students of one member of the team are being successful, that teacher shares the instructional practices that he or she has implemented. Effective instructional practices are often found in articles, professional organizations, or on the Internet (see, for example, http://www.allthingsplc.info).

A culture of experimentation and continuous improvement

Professional learning communities place an emphasis on action, on *doing*. Members of a professional learning community try things out, experiment, and, of course, collaboratively analyze the results of their efforts. By constantly seeking new and better ways of doing things, by trying them out, and by analyzing the effectiveness of their efforts, teacher and student knowledge improves.

A focus on results

In a professional learning community, the primary focus is on *results*—rather than asking "How do you like it?" ask "How has this effort affected student learning?" In such a community, teams of teachers are continually analyzing evidence of student learning, reflecting on the effectiveness of their own practices, and setting meaningful improvement goals. The key to understanding the power of professional learning communities is to understand the power of collaborative teams taking *collective responsibility for results.*

Summary

If the dream of improved student learning in mathematics is to be realized, educational leaders must make deep, substantive changes in the *culture in which teachers teach and students learn*—a change from individual teachers working in isolation while striving to make sure each and every standard is "covered" and taught well, to a collaborative culture in which teams of teachers work to ensure that the standards are, in fact, *learned*—by each and every student.

This change will not be quick, nor will it be easy. Cultural change is a difficult, complex, and incremental endeavor, but it can be done. Many schools like the one described in this chapter currently exist. So, the critical question facing educational leaders is not, "Can it be done?" but rather, "Why not create schools like this for *all* students, and *why not do it now?*"

REFERENCES

Barth, Roland S. *Learning by Heart.* San Francisco: Jossey-Bass, 2001.

Darling-Hammond, Linda. "What Matters Most: A Competent Teacher for Every Child." *Phi Delta Kappan* 78, no. 3 (1996): 193–200.

DuFour, Richard, and Robert Eaker. *Professional Learning Communities at Work: Best Practices for Enhancing Student Achievement.* Bloomington, Ind.: Solution Tree, 1998.

DuFour, Richard, Rebecca DuFour, and Robert Eaker. *Revisiting Professional Learning Communities at Work: New Insights for Improving Schools.* Bloomington, Ind.: Solution Tree, 2008.

DuFour, Richard, Rebecca DuFour, Robert Eaker, and Thomas Many. *Learning by Doing: A Handbook for Professional Learning Communities at Work.* 2nd ed. Bloomington, Ind.: Solution Tree, 2010.

Fullan, Michael. "Professional Learning Communities Writ Large." In *On Common Ground: The Power of Professional Learning Communities,* edited by Richard DuFour, Robert Eaker, and Rebecca DuFour, pp. 209–223. Bloomington, Ind.: Solution Tree, 2005.

Fullan, Michael. *Leadership and Sustainability: System Thinkers in Action.* San Francisco: Jossey-Bass, 2005.

Hargreaves, Andy. "Broader Purpose Calls for Higher Understanding: An Interview with Andy Hargreaves." *Journal of Staff Development* 25, no. 2 (2004): 46–50.

Hattie, John. *Visible Learning: A Synthesis of Over 800 Meta-analyses Relating to Student Achievement.* New York: Routledge, 2009.

Kouzes, James M., and Barry Z. Posner. *A Leader's Legacy.* San Francisco: Jossey-Bass, 2006.

Kruse, Sharon, Karen Seashore Louis, and Anthony Bryk. "Building Professional Learning Communities in Schools." *Issues in Restructuring Schools* 6 (Spring 1994): 3–6.

Marzano, Robert. *What Works in Schools: Translating Research into Action.* Alexandria, Va.: Association for Supervision and Curriculum Development, 2003.

National Council of Supervisors of Mathematics. *The Prime Leadership Framework: Principles and Indicators for Mathematics Education Leaders.* Bloomington, Ind.: Solution Tree, 2008.

National Council of Teachers of Mathematics (NCTM). *Principles and Standards for School Mathematics.* Reston, Va.: NCTM, 2000.

Newmann, Fred M., and Gary G. Wehlage. *Successful School Restructuring: A Report to the Public and Educators by the Center on Organization and Restructuring of Schools.* Madison, Wis.: Center on Organization and Restructuring of Schools, 1995.

Pfeffer, Jeffrey, and Robert I. Sutton. *The Knowing-Doing Gap: How Smart Companies Turn Knowledge into Action.* Boston: Harvard Business School Press, 2000.

Reeves, Douglas B. *The Leader's Guide to Standards: A Blueprint for Educational Equity and Excellence.* San Francisco: Jossey-Bass, 2002.

Reeves, Douglas B. "Putting It All Together: Standards, Assessment, and Accountability in Successful Professional Learning Communities." In *On Common Ground: The Power of Professional Learning Communities,* edited by Richard DuFour, Robert Eaker, and Rebecca DuFour, pp. 45–63. Bloomington, Ind.: Solution Tree, 2005.

Sarason, Seymour B. *Revisiting the Culture of School and the Problem of Change.* New York: Teachers College Press, 1996.

Schmoker, Mike. "Tipping Point: From Feckless Reform to Substantive Instructional Improvement." *Phi Delta Kappan* 85, no. 6 (2004): 424–32.

Sparks, Dennis. "Leading for Transformation in Teaching, Learning, and Relationships." In *On Common Ground: The Power of Professional Learning Communities,* edited by Richard DuFour, Robert Eaker, and Rebecca DuFour, pp. 155–75. Bloomington, Ind.: Solution Tree, 2005.

CHAPTER 2

We're a Learning Community! Now What?

Alan Blankstein

M ANY SCHOOLS and districts throughout North America consider themselves to be *communities of learners*, borrowing a term that became popular following a series of gatherings hosted by the HOPE Foundation beginning in the late 1980s. But what does "community of learners" mean? What does using the phrase "community of learners" mean for mathematics teachers? How does one know if one is part of a true "learning community" versus a community that merely has a new name? This chapter addresses these questions and considers the many possible roles mathematics teachers and leaders can play in a learning community.

If a mathematics teacher really is part of a true learning community, then she or he is likely to communicate with peers and students in ways that reflect new norms and values. Schools have traditionally taught students to find the "right answer," and instruction has focused on the skills of the "three Rs" of education. A community of learners realizes that most problems in life do not lend themselves to a single solution or "right" answer. The new skills called for, both in teaching and in collegial decision making, include the ability to analyze, problem-solve, live with ambiguity, integrate thinking across subject areas toward authentic learning experiences, and to "get along" with one another even in the face of conflicting ideas or conflict-prone people.

What if, for example, the goal of classroom instruction was to get every single student to succeed, and not just determining which ones are able or motivated to do so? And what if the same were true of all teachers and principals within the school or district—that the goal was 100 percent success? How would collegial relations, district-wide supports and structures, and instruction need to shift?

Creating an Atmosphere for Learning

In Rogene Worley Middle School in Mansfield, Texas, eighth-grade teacher Elise Buchhorn introduces the Pythagorean theorem by asking her entire class to find the length of the unknown side of a right triangle through discussing it in small groups. Ms. Buchhorn knows that some of her students tend to lead while others are more passive, so she allows the students to choose their own roles: note taker, liaison to the teacher, reporter, and timekeeper. She walks around to each group for about five minutes to check on their progress and understanding of the problem. The culmination of the task requires each small group to report to the whole class on their findings, their process for coming to their conclusion, and any unanswered questions. (Ms. Buchhorn compiles the questions in one big list for the whole class.) Ms. Buchhorn is comfortable asking her class to solve two- and three-step problems and to justify their answers because she knows her students began solving two-step problems in second grade. She works in a school district where both the elementary and secondary mathematics coordinators share a consistent vision for developing mathematical understandings and practices around fluency and multiple representations (Driscoll 1999).

The process and activities described in Elise Buchhorn's classroom foster a relaxed and engaging atmosphere, one that is well received by all students, even reluctant learners. The goal is clearly not to find out who can or can't solve the problem, but rather to make certain *all* students succeed, participate, and connect to and learn from one another. At the same time, Ms. Buchhorn's goal is to gather valuable information about her students' learning styles and the content areas that require additional attention. Each year, Ms. Buchhorn maintains these same practices to find out how her students (those who have been in the district and school previously and those new to the district and school) learn and which areas require additional attention.

Before further clarifying how teaching, collegial relations, and roles shift within a learning community, it is helpful to have a common framework, system, and definition for discussing a learning community. While there are many definitions, the following description comes from *Failure Is Not an Option: Six Principles for Making Student Success the ONLY Option* (Blankstein 2010).

What Is a Learning Community?

Extensive research exists to support the six principles outlined below for creating and sustaining a high-performing school community (CCSSO 2008; Dufour and Eaker 1998; Kruse, Seashore Louis, and Bryk 1994; Newmann and Wehlage 1995). As a brief overview, the Six Principles are:

Principle 1: Common Mission, Vision, Values, and Goals. Consistency in daily activities, policies, processes, and priorities across schools helps to create a common, coherent focus for the organization.

Principle 2: Ensuring Achievement for ALL Students: Systems for Prevention and Intervention. An improvement plan for every student is in place, and it includes systems for quickly identifying those students in need of additional support. A continuum of support and targeted strategies to help low achievers are developed. The support and strategies make clear that it is the shared involvement and responsibility of all educators (teachers, learning support teachers, and administrators) in the school to assist the low achievers in achieving their goals.

Principle 3: Collaborative Teaming Focused on Teaching and Learning. The faculty and staff view themselves as a community of learners in the common pursuit of improving teaching and learning, with the result being improved achievement for every student. Working interdependently to improve teaching practices is the norm. Structures provide opportunities for collaboration, and processes provide the necessary guidance and support to ensure the system's effectiveness.

Principle 4: Using Data to Guide Decision Making and Continuous Improvement. Decisions are made based on analyzing relevant, timely, and disaggregated data that are consistent over time and collected from multiple sources. School goals, school improvement plans, and the focus of professional development for staff are determined through the collection and analysis of data. Teachers use data to identify the students requiring further support and the best means to support them.

Principle 5: Gaining Active Engagement from Family and Community. The school staff understands the importance of building positive relationships with their students' families. The staff gains a common understanding of conditions that affect the students' learning and the ability of families to support their children so that they can accommodate differences. The school staff finds effective means of reaching out to each family and the community to engage them in their students' learning.

Principle 6: Building Sustainable Leadership Capacity. Leadership extends beyond the purview of the formal school leader and is recognized as a means for ensuring commitment to a compelling long-term vision, no matter what changes or challenges may occur. Informal leadership is valued and strengthened as a means to implement and sustain changes that continually improve student achievement.

Six Principles Working as a System

The Six Principles operate together as a dynamic system. Each interacts and is interdependent with the other. Although the diagram in figure 2.1 suggests a mechanistic relationship between the Principles, they are in fact more interrelated. Just like cogs in a machine, if you push hard enough on one cog, the others will slowly turn. This takes a lot of energy and produces little gain for all the effort. When all Six Principles are operating together, the flow of energy and effort can produce results that will motivate and sustain change in the school's culture.

Each of the Six Principles describes specific actions that, when working together as a system, create an inclusive culture of collaboration focused on ensuring improved achievement for all students. The principles are interdependent and connected. When used to guide our practices, the Six Principles help to create to sustainable learning communities focused on what is meaningful and important to the school or organization. Together, the

Six Principles bring about the thoughtful, smart use of resources for the greatest impact and the best possible outcomes.

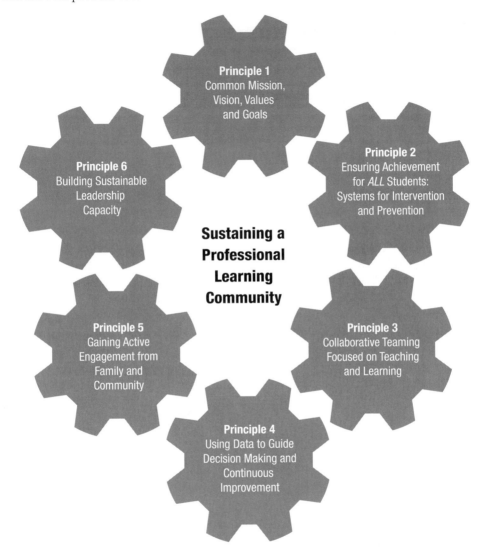

Fig. 2.1. *Failure Is Not an Option* (FNO) System COGS

Let's return to Ms. Buchhorn's class to consider how these six principles work together as a system. Rogene Worley Middle School's mission (Principle 1) of social, emotional, and academic success for every single student is manifest by Ms. Buchhorn, in part by the early assessments (Principle 4) she conducts of each student each year in order to individualize instruction (Principle 2) and, secondly, by her joining with other teachers at her school to introduce new strategies to accommodate new needs (Principle 3). This year, for example, each day Ms. Buchhorn wore a new vocabulary word, such as "irrational," on a card hung around her neck. In addition to helping educate her own students, that card led to new connections with other curious students throughout the school, helping to build overall community. While "irrational" is a term used in mathematics, it could also have allowed her to take the opportunity to reinforce a word that her English-teaching counterparts were emphasizing as part of a cross-curricular literacy initiative led by teachers (Principle 6).

Six Principles Related to S.M.A.R.T. Goals

A learning community sets goals that are *Specific, Measurable, Attainable, Relevant,* and *Time-bound* (S.M.A.R.T.). Setting S.M.A.R.T. goals, a practice first used in the management literature, is a tool for ensuring that the goals can and will be accomplished. In general, here is how the Six Principles would act as a system when used to set and reach a particular S.M.A.R.T. goal.

Specific: The school's mission (Principle 1) helps to identify the goals that the school staff will focus on and use to determine their targeted actions for the year.

Measurable: First, multiple sources of data are collected. Then, through collaborative analysis of the data, the school staff identify the Instructional S.M.A.R.T Goal that targets the specific area needed for student improvement (Principles 1, 3, and 4).

Attainable: Each grade, division, and department identifies the specific instructional strategies and interventions that will be used to support each student. Teachers collaboratively identify appropriate indicators that will provide evidence of student progress toward achieving the level of success required in the Instructional S.M.A.R.T. Goal. They decide on the proper actions for supporting students who are not demonstrating the level and pace of progress identified as appropriate for them. These actions include some that parents may need to take to provide further learning support for their students (Principles 2, 3, 4, and 5).

Relevant: Staff members collaboratively build a professional development plan that will support their own development in using the instructional strategies and interventions identified above. They also build in a plan for providing information, resources, and strategies for parents. The professional development plan includes identifying the appropriate "powerful designs" (Easton 2004) that support their learning, as well as any necessary external support, what evidence will indicate their progress, and a timeline for implementation (Principles 3, 5, and 6).

Time-bound: Teachers collect, examine, and analyze student work to ensure the outcomes being achieved align with the Instructional S.M.A.R.T. Goal. Teachers share this work with the students' parents to inform them of the level of progress being achieved (Principles 3, 4, and 5).

If the evidence indicates students are making progress toward the Instructional S.M.A.R.T Goal, teachers may continue to work to support each other and their own learning as per the professional development plan. If, however, the evidence shows students are not making the required progress, then the professional development plan needs to be reviewed and revised to provide more support for teachers in improving their practice.

Putting the Principles to Work at Mansfield

As an example of a systematic use of the process above, the elementary and secondary coordinators in Mansfield Independent School District (ISD) endeavored along with leadership teams throughout the district to support their mission (Principle 1) by determining ways to more fully engage students' families and students (Principle 5). Each school set its own S.M.A.R.T. goal related to this. At each school a collaborative team of parents and staff (Principle 3), co-chaired by a classroom mathematics teacher and a parent (Principles 5 and 6) began to—

- collect and review the grade-level mathematics curricula;
- collect samples of students' work;
- analyze the areas of need (Principle 4); and
- determine actions to take (Principle 2).

In Mansfield, family engagement was determined to be a priority. The learning community therefore developed several strategies for engaging families in the effort to improve student performance in mathematics. These included enlisting parents as tutors (and training and supporting parents as tutors so that they experienced greater success); having younger students take math games home to their families to teach their elders how to play them; and using the same games at a Math Night with the whole community (Principles 3, 5, and 6).

What Is My Role in Assuring Success?

For both teachers and administrators in a learning community, each person's role shifts away from "I do my job" to "We are collectively committed to the success of all of our students." Understanding the big picture described above and committing to it is the first step. This has implications for approaches to teaching, team membership, and leadership within the school. Learning communities focus on broad goals for students. In mathematics learning, that can include involving students in decision making, problem solving, communication, and assessment (among other things). It is important to focus on these broad goals as you work on content. Each one of these goals is briefly shared below.

Involving Students in Decision Making and Lesson Development

Teachers in Mansfield ISD began to think of more ways to provide more choices for decision making in mathematics learning. Educators should co-create lessons, classroom rules, and even evaluation systems *with* students as much as possible (McTighe 2009). Promoting a learning environment where students can choose how they approach a problem or task also gives students ways to express their content knowledge. These could include how they work on the task (individually or in groups), how they represent or solve a problem (such as drawing a picture, finding a pattern, or experimentation), or how they explain their answer (whether in writing, orally, or in illustrations).

Toni Clarkson, for example, began having her elementary school students demonstrate, with manipulatives and by drawing a picture, two fractions that are equivalent to 2/8. Increasingly, students were asked to explain their thinking orally and in writing. As students worked individually or in pairs, Ms. Clarkson asked questions about the learning goals. For example, as students worked on surface area she asked, "How do a square and cube compare?" Students were asked to write their responses to the question in a journal, which helped them to better articulate their thinking.

Ms. Clarkson also increased creativity into her class. In one example, students were asked to develop a riddle that illustrated a mathematical concept. Inviting children to be creative helps to solidify their mathematical understanding.

Creating Open-ended Problems

In its mathematics instruction, Mansfield ISD also focused on turning closed tasks into open-ended ones. Solving open-ended problems engages both teachers and students as they both think about how to solve the problem and also try to understand other people's ways of solving it. The decision to emphasize open-ended problems had its basis in research and professional development, which learning communities use to inform their work. Groth (2006) describes the major difference between these two problems:

> a. Seven 100-point tests were given during the fall semester. Erika's scores on tests were 76, 82, 82, 79, 85, 25, and 83. Find the mean score.

> b. Seven 100-point tests were given during the fall semester. Erika's scores on the tests were 76, 82, 82, 79, 85, 25, and 83. What grade should Erika receive for the semester?

The first example has one correct answer. The second problem, only slightly altered, is open-ended and engages students in mathematical reasoning. Aside from the various ways an answer could be determined mathematically, the second problem becomes a question of deeper logic. Once the mean score of 73 is determined, some could question why that would be used for an overall grade when all but one of the seven scores are above it (Groth 2006). Moreover, this posing of the problem sparks debate about the purpose of school itself. Why did Erika get a 25? Was it a bad day, a bad test, or poor instruction? Should Erika's entire semester's grade be significantly pulled down by one test score? Is the goal of the school mastery? Teachers and students alike can have a spirited discussion around this question that, if it is allowed, can also help to reinforce the school's mission and values. In Mansfield ISD, teachers collaborated to adapt tasks to open up opportunities for multiple solution strategies and explanations.

Emphasizing Formative over Summative Assessments

Just as with instruction, regularly providing a range of formative assessments provides students with a means of expressing their knowledge (Wiggins and McTighe 2005; Blankstein 2010). Ultimately, all students take some form of standardized test. Leading up to that moment, however, the important thing for the teacher is to know what the student understands, to respond accordingly, and to build the student's ability and confidence in the process. The first example in this chapter demonstrated how Ms. Buchhorn built the collective capacity of her students to problem-solve (using the Pythagorean theorem) and employed an informal "formative" assessment during and at the end of the activity to see how much each student understood. The goal was not to grade or sort students, but to better gauge what they knew so she could plan how to enhance their learning. One problem area she saw had to do with key vocabulary, which is why she daily wore words like "irrational" around her neck. All of Ms. Buchhorn's strategies were driven by frequent, formative, and low-stakes assessments, as opposed to "final" high-stakes exams.

Providing Authentic Learning Experiences

Mathematics is everywhere, in all educational content areas and in our communities and world. Yet mathematics is the content area most often criticized for not being connected to students' lives. As much as possible, mathematics should be connected to other subject areas and made relevant to students' interests. A trip to the national park for a child in a rural farming community could become one that ends in a report that uses science, mathematics, biology, and English to describe and report on the sources of water for the community, the ways in which the community uses water, the quality of the drinking water, and the quantity of water the community uses for irrigation, personal use (such as showers, baths, laundry, or washing dishes), and for drinking.

Effective Collaboration

While many schools now have collaborative team time, it's what happens during that time together that determines ultimate success. Having time set aside to meet is a structural issue; what one discusses in that meeting and does in between meetings is a cultural issue. A small example of this comes from what Ms. Buchhorn learned from Tracey Patton at Worley Middle School during their weekly department meetings. Ms. Buchhorn's students did more poorly than Ms. Patton's in measurement concepts; Ms. Buchhorn's students were performing better in the area of proportionality. Together the two teachers participated in discussions about differences in instructional approaches. The point is that in a true learning community such real, ego-free, and child-centered collaboration frequently occurs and leads to meaningful exchanges and behavioral or teaching strategy changes that yield greater success for students.

True communities of learners have highly productive routines for their team meetings that may include:

- Assuring advance agendas with agreed-upon protocols, and desired outcomes for each meeting.

- Analysis of the data pertaining to each child's performance—disaggregated by race, socioeconomics, gender, and so on. The student data should be tied to individual teachers and classrooms, and it should allow for deeper analysis, such as that Mr. Johnson's students are doing well on quadratic equation questions while Mr. Smith's students are not.

- Commitments to actions by team members based on the collected data. In the above example, one likely action in a high-performing community would be for Mr. Smith to request guidance and feedback from Mr. Johnson on his teaching.

- Follow-through on those commitments. Team evaluation would follow in the next meeting and would pose questions such as: *What was tried? What worked and what did not?* and *What are some recommendations for next steps?*

The important point here is that meeting and talking alone are not sufficient to improve results. A deeper analysis of the data, based on relational trust formed by the team and present throughout the school, is required. Also, the de-privatization of instruction is necessary, so that everyone is learning from one another and best practices are spread throughout the school.

Shared Leadership

The level and the kind of leadership taken by teachers and administrators in a "district of HOPE" can vary from simply open sharing and taking action accordingly (as outlined in the section above in the example of two focused and committed teachers) to organizing learning district-wide. In Mansfield ISD these two types of leadership also happen informally and as part of the formal structure, and one often evolves from the other. For example, Mansfield area administrators Sarah Jandrucko and Jim Vaszauskas spend at least *three and a half days a week* on a campus and in classrooms looking for excellence. Once excellence is found, a teacher may be asked to share it widely. In one recent case, an algebra teacher used a technique that helped his students succeed; when asked about the approach, he suggested organizing a meeting to share it with about fifty other teachers. This in turn led to several days of lesson creation among those same teachers, representing forty schools in the district. In addition, many other teachers indicated their interest in organizing a similar training session spotlighting their own effective practices. Sessions such as these have now become the norm in Mansfield ISD.

Committing to Success

Districts, schools, and individuals that have committed to success and decided that failure is simply not an option are more likely to attain that success. "Failure Is Not an Option" has become the mantra in thousands of schools throughout North America, the United Kingdom, and South Africa. While making such a commitment alone is not sufficient to bring about that outcome, it is a necessary precursor to creating a culture of success. Some of the other essential elements—including a framework and system for action, specific strategies, and changes in roles—are specified above. Collectively, the learning community commits to and acts on the engagement and success of all within it. And that requires changing from the schools we all attended in the past.

We've seen in this chapter the role math teachers and leaders can and do play in creating a high-performing school culture. Some examples of how high-performing schools and districts engage students, collaborate, and effectively lead have been provided. The good news is that much of what is needed to succeed is already present within your school and district: the answer is in the room. Hopefully this chapter will help to identify that answer and spread it throughout your learning community!

......

This chapter is dedicated to George Lenchner, who was the founder of the Math Olympiads for Elementary and Middle Schools (MOEMS) in 1978, author of *Creative Problem Solving in School Mathematics* (1983), and late great-uncle to the chapter's author. The chapter was made possible in great part through the willingness to share information of some extraordinary professionals from the Mansfield ISD, Texas, one of our partner districts. Specifically, helpful conversations were held with Barbara Day, Secondary Math Coordinator; Toni Clarkson, Elementary Math Coordinator; the dedicated teachers Elise Buchhorn, Tracey Patton, and Toni Clarkson; Jim Vaszauskas, Associate Superintendent of Curriculum, Instruction, and Accountability; Sarah Jandrucko, the area superintendent; and Bob Morrison, the district superintendent.

REFERENCES

Blankstein, Alan M. *Failure Is Not an Option: Six Principles for Making Student Success the ONLY Option.* Thousand Oaks, Calif.: Corwin, 2010.

Council of Chief State School Officers (CCSSO). *Educational Leadership Policy Standards: ISLLC 2008.* Washington, D.C.: CCSSO, 2008.

Driscoll, Mark. *Fostering Algebraic Thinking: A Guide for Teachers, Grades 6–10.* Portsmouth, N.H.: Heinemann, 1999.

DuFour, Richard, and Robert Eaker. *Professional Learning Communities at Work: Best Practices for Enhancing Student Achievement.* Bloomington, Ind.: Solution Tree, 1998.

Easton, Lois Brown, ed. *Powerful Designs for Professional Learning.* Oxford, Ohio: National Staff Development Council, 2004.

Groth, Randall E. "An Exploration of Students' Statistical Thinking." *Teaching Statistics* 28, no. 1 (2006): 17–21.

Kruse, Sharon, Karen Seashore Louis, and Anthony Bryk. *Building Professional Community in Schools.* Madison, Wis.: Center on Organization and Restructuring of Schools, 1994.

McTighe, Jay. *Failure Is Not an Option, DVD 3: Effective Assessment for Effective Learning.* HOPE Foundation, 2009.

Newmann, Fred M., and Gary G. Wehlage. *Successful School Restructuring: A Report to the Public and Educators by the Center on Organization and Restructuring of Schools.* Madison, Wis.: Center on Organization and Restructuring of Schools, 1995.

Wiggins, Grant, and Jay McTighe. *Understanding by Design: Expanded Second Edition.* Alexandria, Va.: Association for Supervision and Curriculum Development, 2005.

CHAPTER 3

Closing the Achievement Gap

Systemic Collaboration for Equity in Mathematics

George Ashline
Marny Frantz
Kendra Gorton
Sandra Hepp
Stephanie Ratmeyer

M ILTON ELEMENTARY SCHOOL (MES), a K–6 suburban public school of approximately one thousand students situated in an economically diverse community of ten thousand residents in northwest Vermont, has long grappled with how best to help all its students succeed in mathematics. High administrator and teacher turnover, an emphasis on literacy instruction, and no curriculum for mathematics other than standards-based textbook materials (which teachers had little training for) resulted in inconsistent instruction and expectations for student achievement in mathematics. One of the largest elementary schools in the state, MES serves a population with increasing poverty. Between 2002 and 2008, the percentage of students receiving free and reduced lunch grew from 25 percent to 32 percent.

The catalyst for significant, sustained improvement in mathematics achievement at the school was participation in the Vermont Mathematics Partnership (VMP), a Math and Science Partnership funded by the National Science Foundation and the U.S. Education Department. The VMP was established in late 2002 and comprised seven Vermont school districts, several local institutions of higher education (the University of Vermont through the Vermont Mathematics Initiative graduate program, Saint Michael's College, Castleton State College, and Norwich University), and the Vermont Institutes (an educational nonprofit and the lead partner on the project). VMP has featured mathematicians and education specialists working with school-based leaders to implement sustainable, systemic reform of K–12 mathematics teaching and learning. With deep roots in statewide mathematics education reform efforts, VMP extended the University of Vermont's

Vermont Mathematics Initiative (VMI), a highly successful, content-intensive graduate mathematics teacher leadership program established in 1999. (Among the earlier programs affecting mathematics education: In the 1980s and 1990s Vermont participated in the standards-based reform and authentic assessment initiatives, including Vermont Portfolio Assessment, New Standards Project, and New American Schools. In the 1990s Vermont had a National Science Foundation–funded Statewide Systemic Initiative focused on standards-based curricula and on the development of a state framework of standards and statewide assessments. In the mid to late 1990s Vermont required Action Planning based on student performance.)

VMP work at MES began in the early spring of 2003 with an assessment of the school's strengths and needs in mathematics instruction. Facilitated by VMP's project director, this process included interviews with classroom teachers, specialists, administrators, parents and students, along with classroom observations and a review of instructional materials. With VMP support, the school used this assessment to develop a framework for shared leadership of mathematics teaching and learning, embedded professional development, use of student data as a basis for decision making, and intervention strategies to identify and support students who struggle with mathematics.

One indication that these reform efforts had considerable impact was the improvements at MES in student mathematics achievement on statewide testing. In the late 1990s and early 2000s, MES students scored below state levels in mathematical skills, concepts, and problem solving. By 2004, after only one year with VMP, the percentage of MES fourth graders meeting and exceeding the standard significantly increased, equaling or exceeding statewide results. The growth in problem solving was the most dramatic, going from 24 percent proficient in 2001 to 67 percent proficient in 2004. For comparison, in that same period the statewide growth in problem-solving proficiency went from 31 percent to 52 percent. New statewide standardized tests were introduced in 2005, and progress at MES continued (fig. 3.1). From 2005 to 2008, the percentage of MES students in grades 3–6 who scored proficient or higher in mathematics increased from 65 percent to 69 percent. Most striking is the percentage increase of MES students who scored proficient or higher and also received free and reduced lunch. Only 33 percent of these students met or surpassed the standard in 2005; 48 percent of them did in 2008. Moreover, in that period the achievement gap between MES students who did and did not receive free and reduced lunch narrowed by 11 percentage points; statewide the gap did not change. (*Note:* The New Standards Reference Exam [NSRE], used in Vermont and at Milton Elementary School until 2004, and the New England Common Assessment Program [NECAP] are not aligned. Therefore, it is not possible to track changes in student achievement from the beginning of the project.)

In this chapter, we will describe the partnership's framework (fig. 3.2) for systemic collaboration to improve mathematics teaching and learning at MES. We will present examples of how establishing a dynamic learning community focused on student outcomes led to fundamental changes in mathematics instruction, a narrowing of the mathematics achievement gap, and increases in student achievement across grades K–6. This framework, so effective for mathematics, also became the school's model for reform in science and literacy.

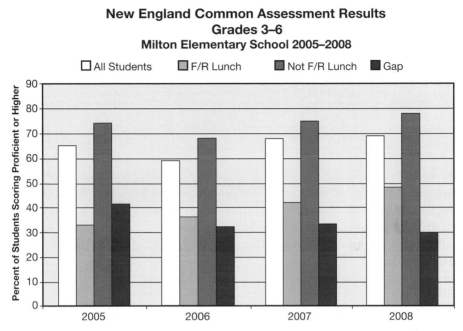

New England Common Assessment Results
Grades 3–6
Milton Elementary School 2005–2008

☐ All Students ▨ F/R Lunch ■ Not F/R Lunch ■ Gap

Fig. 3.1. Closing the mathematics achievement gap at Milton Elementary School

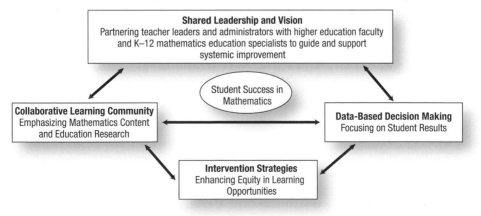

Framework of Systemic Collaboration for Equity in Mathematics
Milton Elementary School and the Vermont Mathematics Partnership

Shared Leadership and Vision
Partnering teacher leaders and administrators with higher education faculty and K–12 mathematics education specialists to guide and support systemic improvement

Student Success in Mathematics

Collaborative Learning Community
Emphasizing Mathematics Content and Education Research

Data-Based Decision Making
Focusing on Student Results

Intervention Strategies
Enhancing Equity in Learning Opportunities

Fig. 3.2. Framework for collaboration between Milton Elementary School and the Vermont Mathematics Partnership

The Framework

One fundamental element of student mathematics success at MES has been intentional, structured collaborations between educators at MES and a group of administrators, higher-education mathematicians, and consulting K–12 mathematics education specialists. ("Educator," when used in this chapter, refers to teachers, special educators, and paraeducators.) These collaborations feature four components:

1. *Shared leadership and vision*
2. *A collaborative learning community* centered on mathematics content and education research
3. *Data-based decision making*
4. *Intervention strategies* to improve equity in learning opportunities

While each framework component plays an important role in enhancing mathematics teaching and learning, our primary focus below will be data-based decision making.

Shared Leadership and Vision

Effective instructional leadership is student-centered and time-intensive, and it requires pedagogical content knowledge combined with strong communication, planning, and interpersonal skills (Marzano 2003; Waters, Marzano, and McNulty 2003). A system for shared mathematics leadership evolved at MES. First, school-based math teacher leaders with extensive training in math content and teaching worked with administrators to establish expectations, and then to implement meeting schedules and protocols for teacher collaboration to improve student math learning outcomes. This school leadership team then partnered with local higher-education mathematicians and external VMP mathematics education specialists to review research and student data, set goals, make plans, offer professional development, and improve the alignment of curriculum, instruction, and assessment.

A Collaborative Learning Community

The second component of this framework features broad participation by faculty and staff in workshops, team meetings, in-service training, classroom coaching, and graduate-level mathematics courses emphasizing the interdependence of rigorous content and research-based pedagogy. An essential foundation for a successful learning community is immersion in significant mathematical concepts along with instructional practices that accentuate clear learning goals, formative assessment, and collaboration (Ball, Thames, and Phelps 2008; Shulman 1987; Conference Board of the Mathematical Sciences 2001). Before grade-level and vertical teams at MES could seriously examine their practices and improve outcomes for students, they had to commit to expanding their professional knowledge base.

The initial assessment of strengths and needs at MES highlighted that many educators had never had an opportunity to explore mathematics deeply or see how the content taught at each grade level connected to prior and subsequent concepts. Most were also unaware of education research findings on how students learn mathematics and on their common errors and misconceptions (Kilpatrick, Swafford, and Findell 2001). Moreover, there was little communication within the school between different grade-level teams or even among educators working in the same grade. Many educators had difficulty implementing the math program, and instructional practices emphasized procedures rather than conceptual discovery and connections (Harris and Nolte 2004–2009). As a result, student exposure to mathematics was dependent on the teacher, and students often experienced discontinuity and gaps in important content.

An initial step toward establishing a schoolwide mathematics learning community and enhancing educator content knowledge was the graduate course Introduction to Mathematics as a Second Language. Teachers, special educators, and paraeducators all took a modified version of this first course in the Vermont Mathematics Initiative graduate program; in it, they explored the concept of number as adult learners and collaboratively examined multiple approaches to help students develop understanding of addition, subtraction, multiplication, and division. The course's instructors modeled diverse formative assessment formats and differentiated activities so that everyone, regardless of background, could extend their understanding and make substantive contributions. Many participants described this first course as transformative. It established a learning community with a common mathematical language and sparked a widespread desire for ongoing professional development in mathematics.

> "Too many of our students have perfected getting lost in their classrooms."
>
> —Kendra Gorton
> Site Math Coordinator
> Milton Elementary School

VMP mathematics educators visited classrooms to coach teachers and demonstrate effective instruction, and this generated even more enthusiasm for professional learning. Teachers watched their own students solve complex problems beyond what they had expected was possible. As a result, teachers raised their expectations and weighed instructional decisions more carefully. They began to move away from attributing low performance to the developmental levels or socioeconomic status of their students and to focus instead on strengthening their own expertise.

Data-Based Decision Making

Gathering and analyzing a wide range of data about mathematics instruction and student performance is critical for informed, systematic decision making by a school's leadership, grade-level teams, classroom teachers, and intervention teams (Mid-continent Research for Education and Learning 2003). Grounded in mathematics content and education research, MES educators and administrators learned to use data to plan schoolwide approaches for teaching mathematics, to identify and design strategies for correcting common student misunderstandings, and to modify classroom instruction to meet the learning needs of all students.

Context for Data Analysis

At MES, data-based decision making was embedded in mathematics improvement efforts. The initial assessment of strengths and needs informed decisions about how best to structure leadership and ongoing professional development. With support from higher-education faculty and VMP educators, the math leadership team regularly reviewed student achievement data; this data included statewide assessments, unit pre-post measures, or school common grade-level tests. Working with teacher leaders, instructional teams then analyzed student performance, discussed past instructional approaches, and applied their deepening professional knowledge to enhance instruction, especially in areas where tests showed weaker student understanding. This included comparing the math content of specific textbook units to the Vermont Grade Expectations, which is the Vermont Department of Education's articulation of minimum proficiency at each grade level (Vermont Department of Education 2008a). This process helped teachers to transition from using program materials only as instructional guides to using them as tools to meet clear learning goals. Rather than just moving through units in a textbook, teachers made strategic choices about ways to highlight lessons, introduce content, make connections, and build understanding.

Analyzing student data was a new process for most teachers, and one that many had long resisted, fearing its primary use would be to evaluate individual teacher performance. The new leadership structure and expectations for collaboration brought a schoolwide shift in perspective. Once it was clear that the emphasis was on helping all students improve, educators grew more comfortable reviewing student results and making informed instructional decisions in light of those results.

Formative Assessment

Professional development at MES focused heavily on formative assessment and making instructional decisions based on data about student understanding. MES teachers participated in developing and implementing mathematics-specific formative assessment tools and processes through VMP's Ongoing Assessment Project. Initially, this project focused on fractions, and the responses students made to some items surprised many teachers.

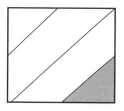

Is ¼ of the square shaded?
Explain your answer using
pictures or words

**Fig. 3.3. Sample VMP ongoing assessment project
item for Grade 3 Pre-Assessment**

Most students who participated in the initial pilot studies responded incorrectly to the item in figure 3.3, and their explanations showed that they either did not consider the size of partitions in the area model or else confused set and area models (Petit, Laird, and Marsden 2010).

Before using pre-instruction probes of this type to gather information about student understanding, teachers had assumed that most students had adequate background knowledge for the instructional units in the textbook materials. The use of formative assessment with fractions highlighted the importance of examining student work for clues about prior understandings or misconceptions. This ultimately led teachers to implement unit pre-assessments for other math concepts and to design appropriately targeted lessons based on the results. For example, a pre-assessment item for a grade 4 unit on place value asked students to add 6 hundreds + 23 tens + 4 ones. Answers typically included such numbers as 834 and 6,234 and 600,234. Depending on the student responses, teachers organized students into small groups to address specific misconceptions and help them extend their understanding.

Prior to implementing formative assessment, teachers generally did not know with any certainty what their students actually learned in a lesson. Now teachers created banks of exit questions on concepts taught at their grade levels. Periodically, they gave exit questions after lessons and sorted results into piles of "got it" or "almost there" or "not there yet," and planned future lessons accordingly. The exit questions showed teachers that just because they "taught it" did not mean that students "learned it." Student responses on these items often disproved teacher assumptions about student conceptual understanding (Ashline 2005).

Formative assessment also helped teachers recognize their need for deeper pedagogical content knowledge to accurately interpret student work and make good instructional decisions. For example, during the fraction items pilot study, teachers took the pre-assessment themselves, and some found that they too held fraction misconceptions, which were being passed on to students. As an illustration, a few teachers believed that 2/3 and 3/4 were "basically the same," because each numerator is one less than its denominator. To help correct such misconceptions and improve their backgrounds, teachers enhanced their own content knowledge in light of current education research regarding student learning of fractions. Pre-assessments and exit questions then aided the teachers in addressing their students' misconceptions.

Common Assessments

The leadership team also established schoolwide systems to collect and review data about students' mathematics understanding. With the support of VMP assessment experts, MES educators developed common grade-level mathematics assessments linked to the Vermont Grade Expectations. Leadership and grade-level teams studied student responses on common assessments to improve instructional decisions. In one example, only 57 percent of a random sample of first graders correctly answered a place value item that related to content they should have understood well by that time. As a result, teachers focused discussion on making sure everyone on the team thoroughly understood the concept, why it was important, and how it related to what students learn in subsequent grades. Then, teachers redesigned lessons to deepen student understanding and crafted additional questions to elicit student thinking.

Analyzing Teaching Practices: A Case Study About Multiplication

In this setting of systematic professional collaboration and data-based instructional decision making, multiplication quickly surfaced as a focus topic. The math teacher leaders for grades 5–6 reported that their students' lack of conceptual foundation in multiplication made it more difficult for them to acquire higher-level understanding, especially in algebra and geometry. At vertical and grade-level team meetings, teachers discussed how multiplication was taught, and they discovered that most teachers presented multiplication procedurally as they followed the textbook. The math leadership team realized that teachers were mostly working in isolation and had little awareness of how multiplication developed across grades.

The inquiry into teaching multiplication grew deeper. A close inspection of student work during grade-level team meetings and other professional development sessions revealed that the level of success students had in multiplication was connected to their understanding of place value. Many students could use algorithms, but they did not really understand what the numbers meant. For example, in the number 254, some students didn't realize that here "2" represents hundreds, "5" represents tens, and "4" represents ones. Without place value understanding, fifth and sixth graders were unable to estimate answers and judge the reasonableness of computations. Also, such students were not fluent with basic math facts or the properties of addition and multiplication. In order to apply the area model for multiplication, students need to understand these concepts.

Place value instruction was inconsistent across classrooms and often occurred in units separate from multiplication. Addressing this problem became the emphasis of schoolwide professional development. Grades K–2 worked on composing and decomposing number, place value, magnitude, and expanded notation. Grades 3–6 math teacher leaders investigated the area model representation for multi-digit multiplication, which combines algebra and geometry in a powerful visual display of the distributive property (fig. 3.4). They shared this model with their grade-level teams, emphasized its place value relevance, and discussed its connections to partial products and precursor K–2 concepts. Some educators included the area model as one of several methods for teaching multiplication.

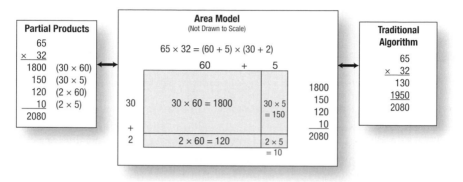

Fig. 3.4. The area model for multiplication with links to other methods

Multiplication issues continued to arise. The math leadership team observed how students solved multi-digit multiplication problems. Despite instructional changes implemented in the previous few years, many fifth and sixth graders did not fundamentally understand multiplication. Grade-level data indicated that students were still trying to memorize steps with no meaning for them and were unable to decide if their answers were reasonable. The data also indicated that some teachers were more successful at helping students understand multiplication; these teachers emphasized conceptual understanding and more effectively used the area model in their lessons.

To enhance their own multiplicative reasoning and investigate models for teaching multiplication conceptually, the math leadership team studied *Young Mathematicians at Work: Multiplication and Division* (Fosnot and Dolk 2001). As a result, this team recommended that teachers in grades 3–4 emphasize strategies that develop conceptual multiplicative understanding, and they suggested postponing teaching the traditional algorithm until grade 5. The team continued to help teachers craft an instructional plan to give students enough time to build a strong foundation, make direct algebraic and geometric links between the area model and partial products, and support an extension from single- to multi-digit problems like 65×32.

Grades 3–6 teachers agreed to try the new approach and requested additional professional development on the area model. Teacher leaders discussed questioning for place value understanding and providing students with opportunities to reinforce K–2 skills. Leaders also presented new instructional strategies that showed connections between the area model, partial products, and the traditional algorithm. Strategies included representing multiplication problems with base ten blocks and using graph paper to make area model arrays that students could separate into parts that were easier to multiply.

"In the past students learned the traditional model procedurally. In the problem 65 x 32 the kids just multiplied single-digit numbers . . . 2 x 5, 2 x 6. They sometimes remembered to put the zero in as a place-holder, then multiplied 3 x 5 and 3 x 6 and got this big answer. They didn't think about place value where the 6 really represented 60.

Now the students are making connections be-tween either partial product or area model and the traditional method. They are able to show how the (30 x 60) + (30 x 5) is the same as the second line in the traditional method.

I have learned that mathematics concepts are not taught in isolation; they are intermingled. It was eye-opening to learn you can teach geometry and algebra concepts through the area model."

—Catherine Thibault-Cote
Grade 5 math teacher
Milton Elementary School

Through studying multiplication and collaborating on instructional improvements, MES educators better understood the content they taught. Table 3.1 shows grades 5–6 scores on the Number and Operations strand of the statewide tests; it is a subset of the statewide testing results shown in figure 3.1. These scores suggest that teachers had successfully translated their knowledge into student learning, especially for students receiving free and reduced lunch. For example, the percentage of these fifth graders meeting the standard in this strand increased from 28 percent in 2005 to 55 percent in 2008. Furthermore, cohorts of these students showed considerable gains in two of three years.

Table 3.1

Subset of Vermont statewide testing results for Milton Elementary School

	Milton Elementary School NECAP 2005–2008, Number & Operations			
Grade 5 (% Meeting Standard)	**2005 (N=122)**	**2006 (N=148)**	**2007 (N=136)**	**2008 (N=130)**
All Students	66.4% (N=81)	71% (N=105)	71.3% (N=97)	69% (N=89)
SES Students	28% (N=7)	54% (N=23)	39% (N=10)	55% (N=17)
Non-SES Students	76.3% (N=74)	78% (N=82)	79% (N=87)	73% (N=72)

Table 3.1—*Continued*

Grade 6 (% Meeting Standard)	Milton Elementary School NECAP 2005–2008, Number & Operations			
	2005 (*N*=133)	2006 (*N*=126)	2007 (*N*=156)	2008 (*N*=134)
All Students	64% (*N*=84)	72% (*N*=90)	72.4% (*N*=113)	78.4% (*N*=105)
SES Students	40% (*N*=10)	48.4% (*N*=15)	55.8% (*N*=24)	66.7% (*N*=16)
Non-SES Students	68.5% (*N*=74)	79% (*N*=75)	78.8% (*N*=89)	80.9% (*N*=89)

Intervention Strategies

The fourth part of the framework is composed of intervention strategies designed to enhance equity in learning opportunities and to supplement classroom instruction for students who struggle with fundamental math ideas. Such interventions target specific curricular content, coordinate with classroom instruction, and involve frequent progress monitoring (Baker, Gersten, and Lee 2002; Bryant et al. 2008).

As the MES math leadership team studied schoolwide student data to understand the difficulties of their struggling students, they realized some students needed additional systemic supplemental support. For example, even when educators recognized that some students did not fully grasp place value, they did not have sufficient time to provide additional support. To address the need for intervention, math leaders first researched effective models. Teacher leaders visited VMP partner Barre City Elementary School's math intervention center and also traveled to New Zealand for discussions and classroom observations. Based on their findings, they developed the Milton Math Intervention Program.

The Milton Math Intervention Program offered intensive, eight-week supplemental support in the school's math lab during the school day for students in grades 2–4. The program assisted students who were not meeting achievement standards, but who were proficient enough that specialized numeracy instruction could help close their gaps in understanding essential math concepts. To determine placement and measure intervention effects, all students at these grade levels took a numeracy pre-assessment at the start of the academic year and an equivalent post-assessment at the end of the year. Regression discontinuity analysis for two years showed that the intervention had a statistically significant impact on post-program test scores. Line plots show the reduction of the gap between mean pre- and post-scores for the intervention and control (non-intervention) groups. The two-year mean gap on the grade 2 pre-score was 30.3 percent, reduced to only 6.3 percent at the post-test, which indicates that the intervention group was functioning very close to the non-intervention group by the end of the academic year (fig. 3.5, Rosenfeld 2009). Further student data analysis suggests that this intervention program may have contributed to narrowing the achievement gap at MES on the statewide assessment (Meyers 2009).

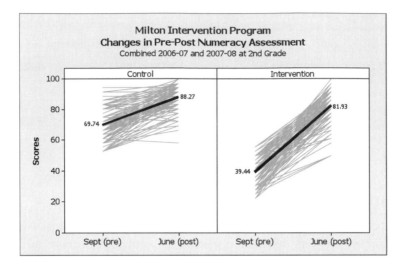

Fig. 3.5. Impact of intervention program on student mathematics performance

The instruction in the intervention program targeted specific student needs in each group. For example, in an eight-week session for second graders, only 13 percent of students identified for intervention initially responded correctly to the following assessment item: $\Box - 5 = 9$. The typical response was "4" or "0." Although other interpretations of this result are possible, the intervention staff concluded that these students lacked an understanding of equality. Subsequent small group instruction focused on explorations and activities with a balance scale, equality sorts, and the meaning of equality. On the post-assessment, given several months later, 87 percent of the students answered this same question correctly.

This intervention model went beyond direct work with students. The ultimate goal was to ensure that *all* MES students had an equal opportunity to learn mathematics. Intervention staff supported students during regular math classes, mentored classroom teachers, and met regularly with math teacher leaders to address schoolwide math issues they had identified through work with students in the intervention program.

Impact of Systemic Collaboration

As shown above, MES students made important mathematics achievement gains after the 2003 introduction of systemic collaboration by educators within and outside MES. Participation in VMP also brought tangible, if less quantifiable, changes to the school. In response to a spring 2007 survey, educators wrote that participation strengthened the school's learning community through a newly shared language for mathematics, regular meetings dedicated to mathematics instruction, and opportunities for professional development across grades, leading to deeper understanding of mathematics content and more effective ways to teach diverse students. The changes educators identified as most important included distributed leadership, effective communication and collaboration, consistent instruction and assessment, and more support for all students, especially those struggling with mathematics (Harris and Nolte 2004–2009).

> "We have become a math community. We are now able to discuss math curriculum and content and learn from each other. A forum is now in place which encourages growth."
>
> —Milton Elementary School teacher, 2007

Ongoing Challenges

MES educators remain committed to the goal of helping all children succeed in mathematics. Ongoing challenges include supporting new teachers in the system, turn-over in administration, balancing competing initiatives within the school, and responding to state and federal mandates. Despite having made reductions in the achievement gap, the school was identified in 2008 as not meeting annual yearly progress requirements for special education students and students who receive free and reduced lunch rates. It is worth noting that such challenges are not unique to MES, but are faced by elementary schools across the country. MES continues to develop an intervention model that emphasizes strong classroom instruction for all students and support for teachers and students identified as needing additional content focus.

Lessons Learned

The framework for systemic collaboration was key to building and sustaining improvement in mathematics teaching and learning at MES. Although we have focused here on data-based decision making, each element of this framework was important, and no single component was enough by itself. A leadership team of educators and administrators from both inside and outside the system articulated a shared vision and established structures designed to support ongoing professional learning, collaboration, data-based decision making, and successful new instructional approaches. Although it was important to re-visit goals and assess progress regularly, the change process we have described was neither linear nor necessarily cyclical. Rather, the framework elements were interdependent. For example, collaborative learning depended on time and access to high-quality professional development. Data-based decision making required the collection of appropriate data, time to analyze it, the knowledge to interpret it, and the authority to make changes.

As the example of how teaching multiplication evolved over several years shows, this dynamic framework not only supported the process of identifying issues and exploring effective ways to address them, but it also helped foster respect and trust among individuals working in the system. Through enthusiastic and respectful inquiry, the entire staff engaged in this process, empowering teachers to set a new course for mathematics instruction. Extensive and ongoing professional development in mathematics content and pedagogy for all educators proved essential. Structural changes were implemented; the entire faculty participated in courses, workshops, or other professional development; and the expertise of teacher leaders impacted the school as a whole. To varying degrees, everyone was involved in these collaborative efforts.

Systemic change took time and persistence. External VMP support helped MES establish a foundation to develop and sustain its ongoing effort. Inevitably, as one issue was resolved, new ones demanded attention. As described earlier, a new leadership structure and an emerging culture of collaboration and data-based decision making enabled MES to respond to the new challenges and help more students succeed in mathematics.

......

The Vermont Mathematics Partnership is funded by a grant provided by the U.S. Department of Education (Award Number S366A020002) and the National Science Foundation (Award Number EHR-0227057). Any opinions, findings, and conclusions or recommendations expressed in this chapter are those of the authors and do not necessarily reflect the views of the National Science Foundation. Kenneth Gross of the University of Vermont is Principal Investigator; Doug Harris and Regina Quinn serve as Co-Principal Investigators.

REFERENCES

Ashline, George. "Integrating Exit Questions into Instruction." *NCTM News Bulletin* 42, no. 3 (2005): 6.

Baker, Scott, Russell Gersten, and Dae-Sik Lee. "A Synthesis of Empirical Research on Teaching Mathematics to Low-achieving Students." *The Elementary School Journal* 103, no. 1 (2002): 51–73.

Ball, Deborah, Mark Thames, and Geoffrey Phelps. "Content Knowledge for Teaching: What Makes It Special?" *Journal of Teacher Education* 59, no. 5 (2008): 389–407.

Bryant, Diane, Brian Bryant, Russell Gersten, Nancy Scammacca, and Melissa Chavez. "Mathematics Intervention for First- and Second-grade Students with Mathematics Difficulties: The Effects of Tier 2 Intervention Delivered as Booster Lessons." *Remedial and Special Education* 29, no. 1 (2008): 20–32.

Conference Board of the Mathematical Sciences. *The Mathematical Education of Teachers.* Washington, D.C.: American Mathematical Society and Mathematical Association of America, 2001. http://www.cbmsweb.org/MET_Document/.

Fosnot, Catherine Twomey, and Maarten Dolk. *Young Mathematicians at Work: Constructing Multiplication and Division.* Portsmouth, N.H.: Heinemann, 2001.

Harris, Douglas, and Penelope Nolte, VMP Evaluation Reports and Evaluation Technical Reports, unpublished. 2004–2009.

Kilpatrick, Jeremy, Jane Swafford, and Bradford Findell, eds. *Adding It Up: Helping Children Learn Mathematics.* Washington, D.C.: National Academies Press, 2001.

Marzano, Robert J. *What Works in Schools: Translating Research Into Action.* Alexandria, Va.: Association for Supervision and Curriculum Development, 2003.

Meyers, Herman. Milton Intervention NECAP Analysis June 2009. Unpublished report. 2009.

Mid-continent Research for Education and Learning. *Leadership Folio Series: Sustaining School Improvement: Data-driven Decision Making.* Aurora, Colo.: Mid-continent Research for Education and Learning, 2003.

New England Common Assessment Program (NECAP) Student Performance Assessment Results (last updated January 28, 2009). http://education.vermont.gov/new/html/pgm_assessment/performance/necap_public_schools_A_D.html.

New Standards Reference Exam (NSRE) Student Performance Indicator Definitions. http://crs.uvm.edu/schlrpt/perform.htm#25.

Petit, Marjorie M., Robert E. Laird, and Edwin L. Marsden. *A Focus on Fractions: Bringing Research to the Classroom.* New York: Routledge Taylor & Francis Group, 2010.

Rosenfeld, Robert. Analysis of the Milton Intervention Program for the Vermont Mathematics Partnership. Unpublished report. 2009.

Shulman, Lee. "Knowledge and Teaching: Foundations of the New Reform." *Harvard Educational Review* 57, no. 1 (1987): 1–22.

Vermont Department of Education (DoE). Grade Expectations (last updated April 28, 2008). http://education.vermont.gov/new/html/pgm_curriculum/mathematics/gle.html.

——————. 2008 School Report for Milton Elementary School. http://crs.uvm.edu/schlrpt/cfusion/schlrpt08/schltwn.cfm?city=Milton.

——————. NSRE Summary Reports for Milton Elementary School. http://education.vermont.gov/new/html/pgm_assessment/cas/school_summary_reports.html.

Vermont Mathematics Initiative (VMI). http://www.uvm.edu/~vmi/.

Vermont Mathematics Partnership (VMP). Available at http://vermontinstitutes.org/index.php/vmp; including information about VMP Equity Framework and Ongoing Assessment Project, respectively at http://vermontinstitutes.org/index.php/vmp/midlevel and http://vermontinstitutes.org/index.php/vmp/ogap.

Waters, Tim, Robert J. Marzano, and Brian McNulty. *Balanced Leadership: What 30 Years of Research Tells Us about the Effect of Leadership on Student Achievement.* Aurora, Colo.: Mid-continent Research for Education and Learning, 2003.

Part II

Students and Student Learning

How does a focus on student learning impact mathematics teaching and learning?

Actually making students—not just the high-achieving students, but *all* students—one of the primary focal points of a professional learning community can be challenging. For one thing, developing professional learning communities in which practitioners feel empowered to honestly share the difficulties they face in meeting the needs of all their students takes time and requires great trust.

In this part of the book, professional learning communities are introduced that have a distinctive focus on students and student learning as the means to impact mathematics teaching and learning. All these PLCs have something else in common: Each is concerned with utilizing and developing frameworks that centralize students' mathematical knowledge and thinking, and their cultures and lived experiences.

Below are some questions to consider as you read through these four chapters that focus on students and how student learning can impact mathematics teaching and learning.

- What are the keys to making all students, particularly students who may struggle in mathematics, or be marginalized in the mathematics classroom for whatever reason, the central focus of a professional learning community?

- What are some ways the participants in our PLC can become more informed about their students' lives as a means of improving the teaching and learning of mathematics?

- In what way can the cases shared in these chapters help to make all students and their mathematical learning a more explicit focus of our professional learning community

CHAPTER 4

Centering the Teaching of Mathematics on Urban Youth

Learning Together about Our Students and Their Communities

Laurie H. Rubel

O NE PROMISING APPROACH to successful mathematics teaching is culturally relevant pedagogy (Ladson-Billings 1995). Centering the Teaching of Mathematics on Urban Youth (CTMUY) is a teacher learning community for high school teachers in New York City dedicated to developing practices of culturally relevant mathematics pedagogy. This chapter describes an educational approach that characterizes that program's professional collaboration: learning mathematics of community as a community.

Defining Culturally Relevant Mathematics Pedagogy

Culturally relevant teaching (Gutstein et al.; Ladson-Billings 1995) is instruction that (1) emphasizes students' academic success, (2) encourages the development of cultural competence, and (3) enables students to develop critical consciousness. The framework of *culturally relevant mathematics pedagogy*, or CureMap, further defines these ideas with respect to the subject area of mathematics.

CureMap is built around three dimensions of teaching mathematics, as shown in figure 4.1. CureMap's core—as shown in the largest cog pictured—is teaching mathematics for conceptual understanding. This approach to teaching mathematics emphasizes the connections between mathematical concepts, procedures, and facts (Hiebert and Carpenter 1992), instead of focusing strictly on procedural fluency. CureMap's second dimension is the "centering" (Tate 2005) of mathematics instruction on the students' own experiences. Finally, CureMap's third dimension is to develop students' critical consciousness about and with mathematics.

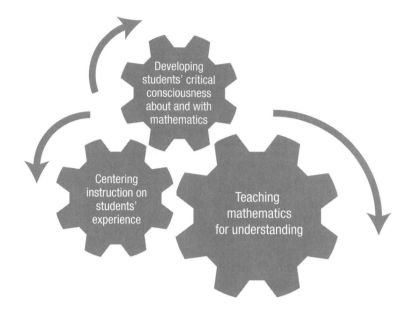

Fig. 4.1. Three dimensions of culturally relevant mathematics pedagogy

While figure 4.1 portrays CureMap's three dimensions as if they are discrete, they actually interact in a variety of ways. For instance, centering mathematics instruction on students is not limited to contextualizing a mathematics lesson within a situation that might be familiar, relevant, or interesting to students. Centering mathematics instruction on students also takes advantage of the ways that classroom instruction can provide opportunities for students to participate in mathematics. In other words, this dimension requires classroom norms and structures that invite and sustain student participation. As a result, students become central participants in the building of mathematical understanding, even if a particular lesson's mathematical content does not have explicit connections to their lives.

CureMap's three dimensions require teachers to have knowledge about their students and the local community. This knowledge helps teachers to create classroom norms that support student participation in building mathematical understanding; to identify relevant contexts for mathematization; and to select (or guide students toward selecting) social issues that can be analyzed or described with mathematics. Learning about students and their communities can be challenging, especially when teachers are outsiders to their students' communities or when a school has a diverse student body, as is typical in urban contexts. The sections that follow present an example of a professional collaboration organized to support high school mathematics teachers in developing practices of CureMap, and more specifically, in gaining knowledge about their students and their students' communities.

Structure of the Collaboration

Centering the Teaching of Mathematics on Urban Youth is a teacher learning community developed around the framework of CureMap. The program is designed to enable high school mathematics teachers to collaboratively engage with mathematics, to reflect on their instructional practices in the light of culturally relevant mathematics pedagogy, and to analyze the relationship between their teaching practices and their students' participation and learning.

I started CTMUY in 2005 by recruiting thirteen high school teachers at ten different schools from among graduates of the Brooklyn College (CUNY) teacher education program, with seed funding from MetroMath Center for Learning and Teaching followed by additional funding from the Knowles Science Teaching Foundation. The 2009–2011 and 2011–2013 CTMUY cohorts are supported by funding from the National Science Foundation, and these cohorts consist primarily of teachers from two partner high schools. This design allows the project to work extensively with two complete mathematics departments at the two partner schools as their primary source of professional development.

The two CTMUY partner high schools are in low-income, urban neighborhoods: one neighborhood is predominately African American and the other, about two miles away, is predominately Latino/a. These schools were selected because they serve students from historically underserved, low-income neighborhoods, struggle with student achievement in mathematics, have administrators and/or mathematics teachers who are graduates of the Brooklyn College (CUNY) teacher education program, and were enthusiastic about participating in this collaborative project.

The CTMUY 2009–2011 learning community consisted of a teacher educator/researcher (this chapter's author), a doctoral student, thirteen teachers and three student teachers from the two partner schools, and six teachers from other neighboring high schools. Among the twenty-two teacher participants, the teachers self-identified their race or ethnicity in the following ways: ten identified themselves as African, African American, or Afro-Caribbean, ten as White, one as Latino/a, and one as European. All the teachers live outside of the neighborhoods in which they teach. At the start of the project, among the nineteen in-service teachers, the most experienced teacher was in her eighth year, and the least experienced was a first-year teacher, with the median number of years of experience at 4.5. Seventeen of the nineteen in-service teachers were inducted into teaching through alternative certification programs, a typical rate for schools in low-income urban neighborhoods.

Project Activities

CTMUY is composed of three types of professional development activities: an annual five-day summer institute, semimonthly or monthly meetings, and facilitator observations of individual teachers' mathematics lessons. The summer institutes focus on improving teachers' culturally relevant mathematics pedagogy through a general theme of learning mathematics of community as a community (a concept described in detail later in this chapter). The summer sessions serve as a common experience for participants and generate momentum for improving practices of CureMap.

School-based learning community meetings, the second type of activity, take place twelve to fifteen times during each of the two school years at each of the two partner schools. Every meeting begins with each participant sharing a "high" and "low" moment or experience related to their mathematics instruction since the previous meeting. This creates a space for colleagues to share reflections and also helps the facilitators to determine which issues are pressing for the group. The remainder of each meeting is devoted to a topic that connects to the CureMap framework. For example, in a meeting focused on teaching for understanding, teachers were introduced to a classification framework for the cognitive demands of mathematics classroom tasks and then practiced rating a variety of these tasks (Smith et al. 2004). In another meeting focusing on the two partner schools' local neighborhoods, we collaboratively examined primary sources of information about these locales, ranging from local history to various descriptive sets of data.

The third CTMUY project activity is one-on-one mentoring sessions with teachers. In my role as project director, I typically visit a teacher in his or her classroom and sit in on lessons at least three times, and up to ten times, each year. Before each visit, the teacher and I communicate by email so that we can collaborate to refine any lesson plans, or to discuss potential mathematical representations or tools. We also typically meet again after the lesson to reflect on how well it allowed for student participation and to look for any potential for improvement. Participants regularly email me (and one another) with questions about an upcoming lesson, to report a particular success, to vent about a frustration, or to ask for suggestions for resources. Teachers in the program also visit each other's classrooms or schools, and they have become important professional resources for one another.

A distinguishing feature of CTMUY is that teachers are not being trained to implement a particular set of lessons or to use a predesignated instructional format. Teachers select the aspects of CureMap's three dimensions they find appropriate for their own teaching. For instance, a first-year teacher might select different challenges than a more experienced teacher. Teachers receive mentoring and support, tailored to their own interests and needs, in the form of individual classroom visits and mentoring sessions, but this occurs within the context of the larger, group program. The many group experiences of the summer institutes, fall seminars, and ongoing learning community meetings offer teachers opportunities to learn from other teachers, at their own school and at neighboring schools. This fosters professional relationships and collaborations that continue beyond participants' formal involvement in CTMUY. Participating teachers regularly make statements that highlight the importance of the group structure, like "I feel really lucky to be a part of this group of teachers and this project" or "It was an incredible experience for me not only because of the activities you chose … but because of the people you brought together."

Mathematics and Community

One way that CTMUY collaborates with teachers to improve practices of culturally relevant mathematics pedagogy is by following the principle of *learning mathematics of community as a community*. The participating teachers are high school mathematics teachers who identify with the formal discipline of mathematics. Therefore, teaching mathematics is our common ground; we do mathematics collaboratively, learning as a community, as the primary means of getting to know one another as mathematics teachers. Yet we are teachers of students in a specific urban context, so we also use mathematics as a primary means of studying communities and developing knowledge about our students.

Mathematics as Community

An essential feature of the CTMUY summer institutes is our doing mathematics as a community. Teachers engage with a variety of types of mathematical tasks, often in collaborative groups, and share their findings with the larger group. For instance, in the 2009 summer institute, participants worked on discrete mathematics topics such as graph coloring and its application to scheduling, and the conflicting results of various election systems. In the 2010 summer institute, participants engaged in in-depth problem solving around several related combinatorics tasks and the connections between them, such as the Trains of Thought problem (Benson 2005, p. 156) and the Simplex Lock problem (see http://www2.edc.org/makingmath/mathprojects/simplex/simplex.asp). One participating teacher remarked on how unusual this activity was: "As a math teacher, you never really get to do math with math teachers."

Doing mathematics as a community gives teachers a chance to experience mathematics as learners and then immediately reflect on that learning experience and its implications for their teaching. In some cases, this means that the teachers incorporate some aspect of the activity in their own teaching, such as the particular mathematical task, the group structure, or the strategies for sharing a group's ideas. More broadly, ongoing participation in doing mathematics as a community prompts teachers to think differently about mathematics teaching and learning and about the subject itself. For instance, the deliberate organization of a collaborative mathematical activity gives teachers an opportunity to experience learning mathematics in ways similar to how they may want their students to experience it. Participants can experiment with multiple mathematical representations and make connections among them, as well as with various mathematical concepts. Teachers typically report that these experiences are powerful and compelling evidence for them of the potential benefits of collaborative learning in the classroom.

Every detail of each mathematics activity session serves as an opportunity to model "best practice" with teachers; this may include ways to assign groups, launch a problem, use physical or technological models, or facilitate a discussion with the whole group that enables a sharing of solutions. Sometimes activities are organized so that groups work on a common mathematical task. At other times the groups work in a jigsaw format, or they work on unique tasks that fit together to contribute to a single theme. In the program we also share initial ideas with the larger group in different formats, and we use different structures to share more finished products across groups. When teachers experience this variety of structures for participation as learners, they become more able to incorporate aspects of these structures into their own instruction.

The summer institutes also include regular sessions for reflecting on how we did mathematics together and how we shared our thinking with one another. We talk about each particular collaborative structure, and how much it encourages or inhibits participation. For instance, we discuss the way that group members can take on or be positioned as having "high or low status" (Cohen 1994), and the impact this positioning has for the individual and the entire group. Diversifying and analyzing participation structures is an important component of every CTMUY session.

Mathematics of Community

In addition to its central core of teaching mathematics for understanding, CureMap also prompts teachers to "center" mathematics instruction on students' experiences and to develop students' critical consciousness with mathematics. In CTMUY, we work specifically with mathematics to describe and analyze aspects of community as a way to explore these dimensions of CureMap. For instance, at an international scale, we examined a technique that uses the area under a curve to describe and compare income distributions across nations (Staples 2005). At the national scale, we analyzed a data set on death penalty rates with respect to the race of both the defendant and the victim (Yates, Moore, and McCabe 1999). We found that while the overall percentage of white convicted murderers who received capital punishment exceeded that of black convicted murderers, when these categories are disaggregated in terms of the race of the victim as well the trend reverses, in an illustration of Simpson's paradox. At the regional scale, in order to evaluate the "Living Wage" movement, we plotted data on federal and New York state minimum wages over time, modeled that data, and refigured these wages to reflect inflation. We have also focused on a theme of maps, or cartograms, as representations that enable us to explore local geographical communities.

Maps of Communities

Maps are connected to culturally relevant mathematics pedagogy in a variety of ways (see Rubel, Chu, and Shookhoff [2011] for more details about the CTMUY focus on maps and mapping). Most generally, maps can be used to highlight the significance of representation as a tool for building understanding. Maps of a local neighborhood or city can be particularly relevant to students, and teachers can use them as contexts for various types of problem posing. Additionally, maps of local communities can function as tools with which to explore, identify, represent, or challenge statements about local resources as well as local inequities.

Local maps can also be related to a variety of mathematical concepts within the high school curriculum. Students can calculate descriptive statistics using data from the U.S. Census or other government or nonprofit agencies. The concepts of ratio, proportion, scale, transformations, area, or perimeter can all be explored through maps. Maps can also be used to investigate and contrast concepts of centers, such as finding the population center of a specific geographical area and comparing it to that area's land center. Furthermore, maps can be used to create vertex-edge graphs, as a way to explore possible routes, circuits, or minimal paths. Such graphs are useful tools for analyzing issues of access related to facility location.

In CTMUY, we also examine a range of types of local maps from a critical perspective. We practice identifying a map's "point of the view" by considering how it is scaled, according to what variables, and how it is oriented. For instance, in one activity we created our own maps, or cartograms, of the five counties of our city, employing scales other than land area. When we represent the space using a scale like average household size, for example, different counties become larger and others become smaller. We mapped our city when scaled according to different variables, per the teachers' interest, such as human population, average household income, number of homicides, or number of Starbucks branches. One can look at a specific cartogram to investigate a single variable of interest across the city, or one can look at several cartograms to focus, instead, on the particular attributes of a specific area of the city.

Community Walks as Tools

A second way we work with maps relates to learning about our students and their communities. The summer institute included activities designed to prompt teachers to reflect on how familiar they are with their students' daily lives. For instance, in one activity, teachers were asked to consider a student from the previous year's class and imagine a day in his or her life, beginning with the student's home and commute to school. One teacher reflected, "It was interesting when you had us write down a day in the life of our students. I think I kind of stopped and I realized that I actually . . . have some students I really feel like I don't know what they do when they go home, or even how they get to school." This prompted this teacher to think differently about her role as a teacher of students whose lives did not seem to resemble her own, and to strategize about ways she could learn more about her students in the next school year.

In another summer institute activity, pairs of teachers received a street map of a unique Census tract (in densely populated areas like cities, each tract usually covers a small area) in a low-income, predominately Afro-Caribbean neighborhood. Teachers were given two options. They could choose a particular theme ahead of time (such as access to financial institutions, diversity of grocery options, availability of recreation spaces, or indicators of gentrification) and then walk through the tract to investigate that theme. Alternatively, they could walk the tract with an open mind to allow potential issues in that neighborhood to emerge. Teachers shared their findings with the whole group and also explored a variety of electronic resources containing information linked to those Census tracts.

In the 2005–2007 cohort of CTMUY, many teachers adapted aspects of the community walk and mapping activities for student investigations. They found creative ways in their teaching to use maps, or physical aspects of neighborhoods, as contexts for exploring various mathematical concepts. These curricular projects (see Rubel [2010] for some examples) successfully brought aspects of the students' communities into mathematics classrooms and demonstrated the teachers' ability to reformulate lessons to focus on these local contexts.

The more general purpose of the community walk, however, is for teachers to use it or related tools to learn about the communities in which their students live. Although the 2005–2007 cohort practiced this activity over two consecutive summers, and teachers actively took up mapping as a context for curriculum, they still resisted ongoing suggestions to use community walks, or other similar tools, to develop their own knowledge about their students. CTMUY teachers rejected suggestions to shadow a student at school, make a home visit (Moll et al. 1992), visit a neighborhood church, or invite a student to lead teachers on a neighborhood tour. These activities were not part of our collaborative work in 2005–2007, and it became clear that a single community walk experience (even if modified and repeated over consecutive summers) was not a sufficiently convincing or compelling learning experience for teachers. An analysis of why the task of learning about students was challenging for the 2005–2007 CTMUY cohort, along with a teacher's perspective, is presented in Chu and Rubel 2010.

Refining and Extending the Study of Community

My experiences as the facilitator of the 2005–2007 CTMUY cohort prompted me to revise the project so that the community walk activities extended well beyond a one-day experience. In the 2009–2011 CTMUY cohort, we began examining urban neighborhoods using a variety of mapping techniques, data, and technologies. We then explored urban communities on foot to practice identifying aspects of places that could be described or analyzed with mathematics. Instead of beginning our work in the low-income communities in which they teach, with the variety of challenges posed by the teachers' "outsider status," we began the series of community walk activities in a local, affluent neighborhood. Pairs of teachers, along with two to four of their own students, conducted a collaborative "mathematics of community" walk. Teams assigned roles to their members, received a neighborhood map, and proceeded on a unique route through that space by collaborating to answer questions that were received via text message. The submission of a correct answer, also by text message, prompted a new set of directions to the next problem-solving space.

The "mathematics of community" walks included three categories of problem solving. One category uses the local architecture as a context for mathematical problem posing. For example, while looking at the iconic Washington Square Arch, shown in figure 4.2, participants computed the area of the arch's opening. This figure, a rectangle topped by a semicircle, is a recurrent architectural theme in this neighborhood, from the Norman windows on a host of neighborhood churches to "the key" on nearby basketball courts. As the groups proceeded through the neighborhood, they were prompted to notice the prevalence of this mathematical object in the local architecture.

Fig. 4.2. Washington Square Arch

A second category of problem solving focused on analyzing the geometry of urban space. Teams were led to a local park, Father Demo Square, which is, ironically, triangular. Participants were guided to "walk" various features of the triangle, as a way to think about different notions of a triangle's center. This particular park has a circular fountain within its boundaries, and groups had to consider if the fountain is located in the park's "center." This led to a contextualized discussion of different ways to define a triangle's center, and how each definition could lend itself to a different location for the center, depending on the properties of the triangle.

The broader goal of a community walk was to learn an approach with which one can learn about any neighborhood. To achieve this, participants on our mathematics of community walk were given four guiding questions:

1. Who seems to live here?

2. Who seems to work here?

3. What kinds of resources does this neighborhood offer its residents?

4. What seem to be the economic challenges facing this neighborhood?

A third category of problem solving, therefore, focused on data about a neighborhood's residents and quality of life. In one example, participants examined data that contrasted the median household income in that affluent neighborhood with the two low-income neighborhoods represented by the participating schools. Perhaps surprisingly, the felony rate in the affluent neighborhood was nearly double the one in each of the two low-income neighborhoods. The groups, composed of students and teachers, were prompted to interpret these and other data. In this case, the students provided plausible interpretations of the differences between these felony rates and helped their teachers make sense of this surprising finding.

The mathematics of community walk was the first in a series of collaborative teacher learning activities with the goal of developing tools for learning about an urban space and its inhabiting communities. We conducted various versions of community walks in the partner school neighborhoods, with a focus on the same set of guiding questions. Before physically exploring the neighborhood, teachers investigated maps of the local space, read about its history, and learned about the community from panels of students, parents, and community leaders. Groups of students, parents, or community leaders collaborated in the community walks, functioning as resident experts about their neighborhoods.

Faced with the challenge of learning about her students, one participating teacher was spurred to a unusual action. After her participation in the 2009 summer institute, she focused on the country of origin of many of her students and applied to a local travel fund for teachers for a grant to visit the Dominican Republic. She was awarded funding for a summer trip, and in 2010 she visited three Dominican cities, experienced several home stays, and visited several local schools. She indicated that "visiting her (students') previous schools, meeting their teachers, observing their hometowns, and interacting with their family members will provide me with ideas to capitalize on students' interests and subsequently modify my instructional strategies." She shared her experiences with the CTMUY teacher learning community as part of the 2010 summer institute.

Final Thoughts

Culturally relevant mathematics pedagogy offers teachers a broad conceptualization of what it means to be an effective mathematics teacher. Teaching mathematics for understanding, centering the instruction of mathematics on students' experiences, and developing students' critical consciousness about and with mathematics are far-reaching concepts. They require that teachers have a deep and flexible knowledge of mathematics, of how to teach that mathematics, and of their students and their students' communities.

A CTMUY teacher reflected on her developing understanding of CureMap by distinguishing between the task of creating locally themed mathematics lessons and the ongoing process of learning about students and their communities. She explained: "I used to think that it [CureMap] was finding material that would be in the curriculum, or putting material into a curriculum that relates to your students' lives. . . . [Now] I think it's a little bit more than that. It's that, plus the idea that I need to know my students, and if

I know them, I'll know what things affect them, what things they find important, and also the things they don't know that are important, and then try to use math as a way to get to them, and to teach them math, but [also] to teach them more about the world around them." CTMUY demonstrates a particular approach to this type of collaborative professional development: both doing mathematics as community and doing mathematics of community.

......

This material is based upon work supported by the National Science Foundation under grant numbers 0742614, 0333753, and 0119732 and the Knowles Science Teaching Foundation. Any opinions, findings, and conclusions or recommendations expressed in this material are those of the author and do not necessarily reflect the views of the National Science Foundation or of the Knowles Science Teaching Foundation. Special thanks to Stray Boots (www.strayboots.com) for their invaluable assistance with the "mathematics of community" walks.

REFERENCES

Benson, Steve. *Ways to Think about Mathematics: Activities and Investigations for Grades 6–12 Teachers.* Thousand Oaks, Calif.: Corwin Press, 2004.

Chu, Haiwen, and Laurie Rubel. "Learning to Teach Mathematics in Urban High Schools: Unearthing the Layers." *Journal of Urban Mathematics Education* 3, no. 2 (December 2010): 57–76.

Cohen, Elizabeth. *Designing Groupwork: Strategies for the Heterogeneous Classroom.* New York: Teachers College Press, 1994.

Gutstein, Eric, Pauline Lipman, Patricia Hernandez, and Rebecca de los Reyes. "Culturally Relevant Mathematics Teaching in a Mexican American Context." *Journal for Research in Mathematics Education* 28 (1997): 709–37.

Hiebert, James, and Thomas Carpenter. "Learning and Teaching with Understanding." In *Handbook of Research on Mathematics Teaching and Learning,* edited by Douglas A. Grouws, pp. 65–100. New York: Macmillan, 1992.

Ladson-Billings, Gloria. "Toward a Theory of Culturally Relevant Pedagogy." *American Educational Research Journal* 32 (1995): 465–91.

Moll, Luiz, Cathy Amanti, Deborah Neff, and Norma González. "Funds of Knowledge for Teaching: Using a Qualitative Approach to Connect Homes and Classrooms." *Theory Into Practice* 31 (1992): 132–41.

Rubel, Laurie H. "Centering the Teaching of Mathematics on Urban Youth: Equity Pedagogy in Action." In *Mathematics Teaching and Learning in K-12: Equity and Professional Development,* edited by Mary Q. Foote, pp. 25–40. New York: Palgrave Macmillan, 2010.

Rubel, Laurie, Haiwen Chu, and Lauren Shookhoff. "Learning to Map and Mapping to Learn our Students' Worlds." *Mathematics Teacher* 104, no. 8 (April 2011): 586-91.

Smith, Margaret, Mary Kay Stein, Fran Arbaugh, Catherine Brown, and Jennifer Mossgrove. "Characterizing the Cognitive Demands of Mathematical Tasks: A Task-Sorting Activity." In *Professional Development Guidebook for Perspectives on the Teaching of Mathematics: Companion to the Sixty-sixth Yearbook,* pp. 45–72. Reston, Va.: NCTM, 2004.

Staples, Megan. "Integrals and Equity: Not the Most Obvious of Bedfellows." In *Rethinking Mathematics: Teaching Social Justice by the Numbers*, edited by Eric Gutstein and Bob Peterson, pp. 103–6. Milwaukee, Wis.: Rethinking Schools, 2005.

Tate, William. "Race, Retrenchment, and the Reform of School Mathematics." In *Rethinking Mathematics: Teaching Social Justice by the Numbers,* edited by Eric Gutstein and Bob Peterson, pp. 31–40. Milwaukee, Wis.: Rethinking Schools, 2005.

Yates, Daniel, David Moore, and George McCabe. *The Practice of Statistics.* New York: W.H. Freeman, 1999.

CHAPTER 5

An Urban Partnership Rooted in Putting Principles of a Formative Assessment System into Practice

DeAnn Huinker
Henry Kranendonk
Janis Freckmann

This year, I gained a deeper understanding of how to use formative assessment principles to guide my teaching. In particular, I am now able to better communicate with my students by providing more descriptive feedback, both oral and written. There was a lot more discussion with students on their progress, misconceptions were cleared up more readily, and students were able to deepen their understanding of the math. Students also seemed to have more purpose in their work.

—School-based math teacher leader

A BODY OF EVIDENCE suggests that we need to be more purposeful in our use of formative assessment practices. A seminal research review by Black and Wiliam (1998) found that an effective use of formative assessment resulted in profound achievement gains for all students, with the largest gains for the lowest achievers. The National Mathematics Advisory Panel (2008) has recommended that formative assessment be an integral component of instructional practice in mathematics. According to the panel, formative assessment leads to increased precision in how instructional time is used during classroom teaching and assists in identifying specific focus topics and instructional needs.

What does a regular use of formative assessment look like in practice? How does its use allow teachers to gain insight into important mathematics and student learning?

How can the implementation of formative assessment be supported systemically across a district? In this chapter, we will describe how a large urban school district and university joined together with the shared goal of linking classroom practice in mathematics to principles of formative assessment. We discuss ten principles of formative assessment and give examples of its impact on classroom practice and student learning. In addition, we summarize the lessons learned through this evolution of a district-wide culture for formative assessment, one that was built up over time as people worked together, solved problems, and confronted challenges.

The Partnership

The Milwaukee Public Schools, with approximately 83,000 students, is the largest district in Wisconsin. In this high-minority, high-poverty district, more than 78 percent of students are eligible for free or reduced-price lunch. The University of Wisconsin–Milwaukee is the largest postsecondary institution in the area, and it prepares a majority of teachers for the district. The university and the school district have a long history of collaborating on grant-related professional development projects for mathematics. In fall 2003, this collaboration was strengthened with an award from the National Science Foundation Mathematics and Science Partnership Program for the Milwaukee Mathematics Partnership (MMP). Both partners made a long-term commitment to work together and share in the leadership for student success in mathematics. As members of this partnership, our work has involved regular, nearly weekly meetings and conversations among district mathematics specialists and university mathematicians and mathematics educators.

What has distinguished the MMP from earlier collaborative efforts is that from its beginning the partnership has consciously defined its purpose through a jointly agreed-upon comprehensive mathematics framework. This framework, shown in figure 5.1, lists five components of mathematical proficiency—*understanding, computing, reasoning, applying,* and *engaging* (National Research Council 2002); these elements surround the Wisconsin content standards of *number, algebra, statistics/probability, geometry,* and *measurement.* The framework provides a summary statement of the vision and purpose of the MMP and places rigorous student learning of mathematics at the center of the partnership's work.

This mathematics framework started discussions that led to a remarkable reform effort that has resulted in significant gains in student achievement in mathematics in grades 3–8, along with a narrowing of achievement gaps. It would be difficult to provide a complete history of the MMP here, but its major contributions can be summarized through the following account; it underscores the collaborative effort required to generate district-wide reform through a formative assessment process that reached into the daily classroom experiences of mathematics for students and teachers.

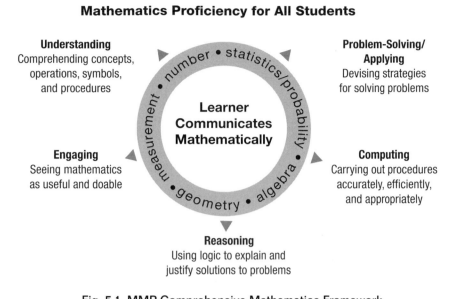

Mathematics Proficiency for All Students

Fig. 5.1. MMP Comprehensive Mathematics Framework

Beginning a District-wide Paradigm Shift in Assessment

After creating the Comprehensive Mathematics Framework, the partnership developed district targets for student learning that were based on state standards and then created classroom assessments (i.e., constructed response items) aligned to these targets. Both of these tasks were a paradigm shift for teachers of mathematics who were accustomed to viewing assessment mainly as the end-of-chapter test.

The development of the targets and assessments was indicative of how this partnership relied on genuine collaboration between university faculty, both mathematicians and mathematics educators, and district teachers. Committees of teachers working with university partners defined the learning targets—statements of the big mathematical ideas students are expected to learn at each grade level, K–8, and at foundational and advanced levels in high school. It was critical that the targets incorporated a conceptual progression of what students needed to learn and when they needed to learn these ideas. This process incorporated the expertise of mathematicians, who saw the underlying mathematical goals, along with the experience of classroom teachers, who provided their knowledge of students and the practical challenges that needed to be addressed to achieve those goals. The result of these discussions was learning targets that teachers valued as workable, that identified progressions of conceptual and procedural understanding and skills, and yet were viewed as attainable by students as demonstrations of mathematical proficiency.

In a similar manner, classroom assessments were developed to support the targets from a perspective of both the mathematics and of what was reasonable in working with students. Again, committees of teachers met with university faculty to write, pilot, review student work, and finalize the classroom assessments. The work began with an intense first year of collaboration in developing assessments for grades 3–5. This was expanded to grades 2–9 the following year and to grades K–12 the year after that. Developing the learning targets and classroom assessments was instrumental in beginning to change the culture in the district from a focus on summative assessment to the use of formative assessment. Teachers, as a result of working with university mathematicians and mathematics educators, were thinking more than ever before about the big mathematical ideas their students were to learn, and now they also had classroom assessments as a tool to monitor that learning.

Structuring Formative Assessment for Practice

As the partnership continued its work, we learned that just developing sets of classroom assessments was not enough. Teachers still wondered what to do with the classroom assessments: When to give them? How to score them? And how to use the results? Initially the assessments were often used as summative measures of student learning, such as being given along with or replacing the chapter test. This was not the intent of the partnership. We had hoped teachers would use the classroom assessments in an ongoing manner to gain formative information on student learning that could be used to guide daily instructional decisions. When we realized this was not happening, we saw the need to increase understanding and give more guidance to teachers in using formative assessment as an integral component of classroom practice.

The partnership worked through these challenges together. Teachers were asked to think about their current assessment practices and the purpose of each practice. Was it to record student progress, to monitor student learning, to guide student learning, or to inform instructional decisions? The partnership knew there had to be movement by teachers toward a balanced classroom assessment system that included both formative assessment and summative assessment. Early discussions revealed that most teachers relied heavily on summative assessments. Homework, quizzes, and tests were given and graded and the scores were recorded; then the class moved on to the next lesson or instructional unit. Of course, some teachers used these assessments to inform and guide instruction, but that was not the usual culture of mathematics assessment throughout the district.

If learning targets and classroom assessments were to make a difference in classroom practice, teachers in the district would need to understand how to use these tools more effectively to impact student learning as part of daily instructional practice. The partnership also needed to better understand how to support teachers through this transition from summative to formative assessment. We studied the work of Paul Black and Dylan Wiliam

(Black and Wiliam 1998; Black et al. 2004), Rick Stiggins and colleagues (Chappuis et al. 2005; Stiggins 2002, 2006; Stiggins et al. 2004), and Shirley Clarke (2001).

Our first insight was the need to distinguish between summative and formative assessment. If we consider an analogy to plants, summative assessment would be the process of simply measuring the plants. It might be interesting to compare and analyze measurements, but, in themselves, these do not affect the growth of the plants. Formative assessment is the equivalent of feeding and watering the plants appropriate to their needs, which directly affects their growth.

Furthermore, Stiggins (2002) provided a useful distinction between "assessments *of learning*" that measure achievement at a point in time for purposes of reporting and "assessments *for learning*" that serve to promote greater student learning (p. 761). This distinction between assessment "of learning" and "for learning" has been useful in our work; it has helped us to develop a comprehensive view of formative assessment as a process that "happens in the classroom and involves students in every aspect of their own assessment" (Stiggins and Chappius 2006, p. 11).

Our second insight was to recognize that we had emphasized only two of the principles of formative assessment we had formulated: alignment of lessons to state and district learning expectations (Principle 4, as shown in fig. 5.2), and use of classroom assessments (Principle 5). We were struck that our perspective had been so narrow, and we realized our need to be more intentional and specific in providing a structure of principles that would support assessment *for learning* across the district. We modified existing lists (Chappuis et al. 2005; Stiggins et al. 2004) and arrived at the set of ten principles of formative assessment in mathematics listed in figure 5.2.

Through the MMP, each school had chosen an individual to serve as its math teacher leader. These individuals had full-time classroom teaching responsibilities with the additional expectation of providing mathematics leadership for their schools. The partnership facilitated professional development sessions on the ten principles for the math teacher leaders and developed tools for them to use with the teachers in their schools. The partnership has made great progress in unpacking these principles and assisting teachers in linking them to classroom practice. However, in many ways, our journey has just begun. In the past, the partnership dabbled at the periphery of assessment practices, mainly increasing our use of constructed response items and wondering what to do with the results. Once we began using these ten principles, we were able to focus more directly on the interactions of teachers and students in classrooms.

What follows is an explanation of the principles with examples of how they have been put into practice. We have grouped them into three areas to emphasize the core purpose of each set of principles and the connections among them: (1) articulation of math learning goals by teachers and students, (2) teacher use of assessments to guide teaching, and (3) student use of assessments to move learning. The ordered list as shown in figure 5.2 presents a general sequence of implementation, but in practice the use of the principles is more interactive and synergistic.

Ten Principles of Formative Assessment

Articulation of Math Learning Goals by Teachers and Students

(1) Prior to teaching, teachers study and can articulate the math concepts students will be learning.

(2) Teachers use student-friendly language to inform students about the math goals they are expected to learn during the lesson.

(3) Students can describe what mathematical ideas they are learning in the lesson.

(4) Teachers can articulate how the math lesson is aligned to district learning targets, state standards, and classroom assessments, and how it fits within the progression of student learning.

Teacher Use of Assessments to Guide Teaching

(5) Teachers use classroom assessments that yield accurate information about student learning of math concepts and skills and use of math processes.

(6) Teachers use assessment information to focus and guide teaching and motivate student learning.

Student Use of Assessments to Move Learning

(7) Feedback given to a student is descriptive, frequent, and timely. It provides insight on a current strength and focuses on one facet of learning for revision linked directly to the intended math objective.

(8) Students actively and regularly use descriptive feedback to improve the quality of their work.

(9) Students study the criteria by which their work will be evaluated by analyzing samples of strong and weak work.

(10) Students keep track of their own learning over time (e.g., journals, portfolios) and communicate with others about what they understand and what areas need improvement.

Fig. 5.2. Principles of formative assessment for mathematics

Principles for Articulation of Math Learning Goals (Principles 1–4)

The first four principles focus on making the mathematics that is being learned and assessed explicit to both teachers and students. Principle 1 states that teachers study, prior to instruction, the mathematical concepts and ideas students are to learn. The purpose of this principle is for teachers to revisit their own understanding of the mathematics they will be teaching, identify the essential ideas students are to learn, and examine how those ideas are developed across a sequence of lessons. Teachers use their mathematics textbooks and other resources to review their understanding of the mathematics and of the pedagogy they will employ. As they study, independently or with colleagues, they clarify expectations for student representations and justifications as evidence of student learning.

Once the math concepts are clarified, the second principle indicates that teachers should articulate, in student-friendly language, the specific math goals for the lesson. For example: "Today we're going to learn about using benchmarks of 0, 1/2, and 1 to help us learn about the size of fractions." The third principle involves students in describing the mathematical ideas they are learning in the lesson. In guiding teachers, the partnership has emphasized the importance of posting the daily goal in writing as a focus for the lesson and to engage the students in considering the intent of the goal (Clarke 2001). A teacher may request, "Students, take a moment to read the math goal for today and restate it in your own words to your partner." Others may discuss it as a whole class: "Please share with the class what it is you think we are going to be learning today."

Sharing the math goals, just for a minute or two, puts a frame around the math lesson. Teachers were initially skeptical of posting their math goals, but they are now reporting a surprising influence on their practice. Many teachers have said that the extra time it takes to consider how to articulate the goals in student-friendly terms made them more aware of the mathematical targets of their lessons. They have also have found it helps to keep themselves and the students more focused on the important mathematical ideas throughout the lesson, as opposed to just working through a series of activities.

For example, teachers keep the math goals of the lessons explicit by asking students to state what they are learning as they work through the activities and share in the discourse of the lesson. A teacher might ask, "You said that 2/5 is close to one-half. This is one of the benchmarks we are learning about today. How do you know when a fraction is close to one-half?" Students were more attentive and engaged, and would refer to the posted goal during the lesson. In fact, students were coming to expect these daily math goals and even asked for them if a teacher began a lesson without explicitly stating or posting the intended math goals.

The fourth principle is a reminder for teachers to examine how a sequence of lessons can move students along a trajectory of learning. This is often accomplished as grade-level groups of teachers meet to examine alignment of their lessons to district learning expectations and state standards, as well as to the anticipated use of the classroom assessments.

This set of principles has created a culture of teachers studying daily classroom lessons from a perspective of being able to explicitly state their mathematical ideas and the development of those ideas. Some of the key questions that arise include: What is the math? How should I phrase this math goal for my students? What do I expect my students to be able to say about the math they are learning? How do these goals align to district and state standards? How do these math ideas build on what they already know? What do I need to do to help my students learn these ideas? Where are students expected to go next in their learning of this math?

Principles for Teacher Use of Assessments to Guide Teaching (*Principles 5 and 6*)

The first set of assessment principles involves preparing to use assessment information to inform and guide teaching. Once the mathematical goals are explicit, the next step is to consider the use of classroom assessments. Principle 5 states that classroom assessments need to yield information about student learning of concepts and skills and the use of mathematical processes. Stiggins (2006) explained that an assessment must be designed to serve a specific and predetermined purpose, be related to expected success criteria, and be sensitive enough to detect and reveal student understanding. The sixth principle reminds teachers that assessment information should provide sufficient detail to focus and guide teaching and to motivate and guide student learning.

As noted earlier, the partnership developed a bank of classroom assessments aligned to the district learning targets for mathematics. One of these assessments is shown in figure 5.3. Each assessment attempts to uncover aspects of students' conceptual understanding and mathematical reasoning, as well as misconceptions. Suggestions for when to use specific classroom assessments were incorporated into the district curriculum pacing guides. Teachers were encouraged to use these assessments to check in on student learning at various points during an instructional unit, such as toward the beginning to gauge initial knowledge, during the unit to check in on student understanding, or toward the end as evidence of learning.

To support implementation of the principles, the partnership developed the Classroom Assessment Summary Report (fig. 5.4) as a tool for teacher collaboration within schools. We encourage teachers to work together as grade-level groups in (1) selecting common classroom assessments to administer at key points during a unit of instruction, (2) explicating the mathematical expectations of the task, (3) analyzing student work from the assessments through filters of expected student reasoning and common misconceptions, and (4) making decisions on the next instructional steps. The form expects teachers to articulate the mathematics their students should demonstrate through the task, and then to summarize what was demonstrated through the assessment and identify an instructional strategy to move that understanding further.

Before teaching a sequence of lessons or an instructional unit, teachers meet as grade-level groups in their schools to select common classroom assessments and identify mathematical expectations. The school's math teacher leader participates in these common planning sessions when possible to facilitate and guide the discussion. The teachers discuss what mathematics (i.e., concepts, skills, or processes) they are targeting and how understanding will be evidenced in the student work in order to yield accurate information on student learning. They then write out these expectations for student learning on the report form. This step is critical because the expectations for students will subsequently guide instruction and the feedback given to students.

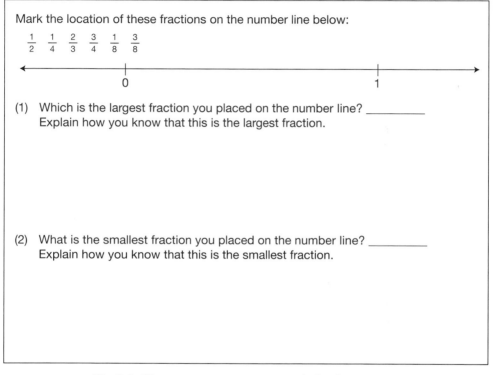

Fig. 5.3. Classroom assessment on ordering fractions

For example, a group of fourth-grade teachers selected common assessments to give at intentional points of an instructional unit on fractions. A teacher reflected:

> In the midst of our fraction unit, we gave a task to assess students' understanding of using benchmark fractions to think about the size of fractions. We had identified the use of benchmarks as a very important idea as students continue developing their ideas on fractions. For the common assessment, we carefully picked some fractions and asked our students to place them on a number line and explain their reasoning for each placement. We then got together to look at the student work and identified areas where our students were being successful and areas in which they were struggling. It really helped me to think this through with my colleagues, and we made some modifications to the next few lessons that we were planning to teach.

Classroom Assessment Summary Report

School:	Date:	Grade Level:
Teacher(s):	___ Regular Education ___ Special Education	

Classroom Assessment: (brief description or attach a copy of the classroom assessment to this report)

State Mathematics Content Standard(s): □ A. Mathematics Processes □ B. Number Operations and Relationships □ C. Measurement □ D. Geometry □ E. Statistics and Probability □ F. Algebraic Relationships **State Math Assessment Framework Descriptor:**	**Mathematical Expectations of Students on the Task:**
Students' Successes in Reasoning:	**Students' Misconceptions and Challenges:**
Next Steps for Instruction:	

Fig. 5.4. Classroom Assessment Summary Report

After administering the common assessment, teachers get together during their biweekly or monthly common planning time to look at the student work from across their classes for evidence of understanding the mathematical expectations. Teachers usually sort the work according to successful reasoning strategies and common misconceptions. They use the report form to help them summarize information about what students understand in light of the expectations and what misconceptions exist. Then they discuss next steps for instruction and modify lesson plans as needed to move students' understanding.

Principles for Student Use of Assessments to Move Learning (Principles 7–10)

It is common practice for teachers to conduct assessments, examine student work, keep records, and track student progress. This third set of formative assessment principles puts a focus on engaging students in using assessment information. When students regularly self-assess, monitor, and communicate their own progress, their confidence in themselves as learners and their motivation to do well grows along with their improved performance (Stiggins 2006).

As the fourth-grade teachers collaboratively analyzed the student work from the classroom assessment on fraction benchmarks, they identified student strengths and misconceptions as the basis for writing descriptive feedback to their students. They noted, "This group of students need to work on knowing when a fraction is greater than or less than one-half, especially when the denominator is an odd number." The teachers then looked together at individual student papers and wrote descriptive feedback on ways the work could be improved. One teacher commented, "I would like to have a whole-class discussion on what makes a good explanation when talking about the size of fractions and the use of benchmarks." They selected two student work samples with correct placement of the fractions but unclear explanations and two samples with clear explanations. After removing the student names, they showed the samples one at a time to the class with the following instructions: "After you read this explanation, restate what this student is saying. What questions would you like to ask this student? What suggestions might you have for making the explanation clearer?" Following the discussion, the students reviewed the written descriptive feedback they had received individually and then revised their work.

Principle 7 states that the feedback given to a student needs to be descriptive, frequent, and timely. It should provide insight on a current strength and identify one facet for revision. The feedback we give to students has more impact than many of us realize. Feedback can help move student learning forward, or it can hinder learning as a result of its negative influence on student motivation (Black et al. 2004; Stiggins 2006). Hattie and Temperly (2007) reported that feedback works to support student learning when it focuses on aspects of the student's work and is descriptive of that work, revealing to the student how to do better the next time (e.g., "Here is how to improve your explanation"), rather than being evaluative or summative (e.g., "Good work," "80 percent"). In order to improve the quality of their work (as noted in Principle 8), students need to have opportunities to use or apply the feedback as soon as possible. This might involve revising the current work or applying the information to a similar task.

This idea of a "do-over" in mathematics was novel to many teachers and was met with some resistance. Common practice in many mathematics classrooms was to give students one chance to complete an assessment and then to record the "score." With the move toward thinking of assessment formatively, more and more teachers across the district are accepting this "do-over" idea. Thus, teachers are giving students feedback that describes an aspect of the work that can be improved, and they are then giving students opportunities to revise and improve their math work. These two principles—descriptive feedback and do-over or revision—have contributed greatly to a cultural shift in mathematics classrooms where more teachers are using assessment for improving student learning rather than just for monitoring or reporting on it.

After learning about the importance of descriptive feedback, many teachers began asking, "Do I still need to score the classroom assessments or is it okay if I just write feedback?" This question itself is evidence of a paradigm shift toward using assessment information to guide learning, rather than placing emphasis on the summative practice of putting a score or final grade on a paper. Butler (1998) found that descriptive feedback by itself is not only sufficient but also desirable, because giving a score or grade can cancel the beneficial effects of the comments. Some teachers in the district now provide only descriptive feedback on the classroom assessments, whereas others provide written comments initially and then provide a rubric score after the student has had an opportunity to revise the work based on the feedback.

Principle 9 says that it is helpful for students to study the criteria by which their work is evaluated. In whole-class discussions, students can analyze samples of strong and weak work to identify characteristics of "good work" and discuss specific ways to improve their work. Through this type of analysis, students deepen their understanding of the mathematics and the attributes that must be represented on the paper to show evidence of learning. This analysis of work samples could occur prior to students reviewing the teacher feedback or prior to self-assessing their own work on established criteria related to the math objectives.

This brings us to the tenth principle of formative assessment, the involvement of students in monitoring their own learning. What students think about and do with assessment information is perhaps more important that what we, as teachers, do with it. When the mathematics learning objectives are clear to students, we can help them learn to engage in self-assessment, goal setting, and identification of their own strengths and areas for improvement. For example, students might summarize in writing what an assessment demonstrates about their understanding and learning needs, and add suggestions for how the teacher or a parent might help them.

Summary

The formation of the Milwaukee Mathematics Partnership provided a structure for ongoing collaboration, but clearly the implementation of the assessment principles generated a genuine working relationship. The partnership was begun in an attempt to bring about a system-wide change in mathematics instruction, but we soon realized that our initial foray into the creation of learning targets and classroom assessments was clearly not enough to move this large district. What was needed was a cultural shift in mathematics classrooms, a shift to formative assessment practices. As one of the school-based math teacher leaders noted:

> Formative assessment was a huge area of growth for me this year that began when we watched a snippet of the Stiggins video at our first Math Teacher Leader meeting this year. I showed the video to my staff, and we began a yearlong journey studying assessments that could be used in the classroom and discussing the need for formative assessment to guide instruction.

With the articulation of these ten principles of formative assessment, our journey is now more focused on ways to link these principles to classroom practice. A cultural shift is occurring in classrooms as we move toward viewing assessment as something that is not done *to students* but rather *with students*. Teachers began by studying the mathematics and translating the lesson math goals into student-friendly language. Students are now expected to be able to articulate the mathematical ideas they are learning. Teachers are expected to know how a math objective fits within the progression and trajectory of student learning. Grade-level groups of teachers now identify and give common classroom assessments and then use the assessment information acquired to make instructional decisions. A huge shift is occurring as we continue the journey and invite students to be more active participants in the assessment process. We are now learning about more ways to provide students with descriptive feedback and give them opportunities to use and apply it. We are also exploring the use of portfolios in the math classroom.

As the partnership has evolved, the mathematics achievement of our students has improved. The district's scores on the state mathematics test have shown a steady increase since the formation of the MMP. The test assesses all state standards and includes both selected response and constructed response items. The percentage of students demonstrating proficiency in mathematics has risen 11.2 percent from 2005 to 2009. In addition, an external evaluation indicated that these gains were most noticeable in schools that were more highly involved in implementing the expectations of the Partnership, such as putting the principles of formative assessment into practice.

The partnership was unlike many previous initiatives of the Milwaukee Public Schools and the University of Wisconsin–Milwaukee, as it was rooted in collaborative learning experiences that were intentionally supporting the principles of formative assessment in a classroom environment. Although far from where we want to be, the partnership has created a culture of collaboration centered on discussions of essential mathematical ideas for student learning grounded in principles of formative assessment, which is a significant change in this urban district.

......

This material is based on work by the Milwaukee Mathematics Partnership (MMP), which is supported by the National Science Foundation under grant number 0314898. Any opinions, findings, and conclusions or recommendations expressed in this material are those of the authors and do not necessarily reflect the views of the Foundation. Additional information on the MMP, including copies of classroom assessments, can be found at www.mmp.uwm.edu.

REFERENCES

Black, Paul, Christine Harrison, Clare Lee, Bethan Marshall, and Dylan Wiliam. "Working Inside the Black Box: Assessment for Learning in the Classroom." *Phi Delta Kappan* 86 (September 2004): 9–21.

Black, Paul, and Dylan Wiliam. "Inside the Black Box: Raising Standards through Classroom Assessment." *Phi Delta Kappan* 80 (October 1998): 139–48.

Butler, Ruth. "Enhancing and Undermining Intrinsic Motivation: The Effects of Task-Involving and Ego-Involving Evaluation on Interest and Performance." *British Journal of Educational Psychology* 58 (February 1998): 1–14.

Chappuis, Stephen, Richard J. Stiggins, Judith A. Arter, and Jan Chappuis. *Assessment for Learning: An Action Guide for School Leaders.* 2nd ed. Portland, Oreg.: Assessment Training Institute, 2005.

Clarke, Shirley. *Unlocking Formative Assessment: Practical Strategies for Enhancing Pupils' Learning in the Primary Classroom.* London: Hodder and Stoughton, 2001.

Hattie, Joan, and Helen Temperly. "The Power of Feedback." *Review of Educational Research* 77 (March 2007): 81–112.

National Mathematics Advisory Panel. *Foundations for Success: The Final Report of the National Mathematics Advisory Panel.* Washington, D.C.: U.S. Department of Education, 2008.

National Research Council. *Helping Children Learn Mathematics.* Washington, D.C.: National Academies Press, 2002.

Stiggins, Richard J. "Assessment Crisis: The Absence of Assessment FOR Learning." *Phi Delta Kappan* 83 (June 2002): 758–65.

————. "Assessment for Learning: A Key to Motivation and Achievement." *Edge* 2 (November/December 2006): 3–19.

Stiggins, Richard J., Judith A. Arter, Jan Chappuis, and Stephen Chappuis. *Classroom Assessment for Student Learning: Doing It Right—Using It Well.* Portland, Oreg.: Assessment Training Institute, 2004.

Stiggins, Rick, and Jan Chappuis. "What a Difference a Word Makes: Assessment FOR Learning Rather than Assessment OF Learning Helps Students Succeed." *Journal of Staff Development* 27 (Winter 2006): 10–14.

CHAPTER **6**

Adapting Instruction for Latina/o Students

A Collaborative Experience between Teachers and Researchers

José María Menéndez
Laura Kondek McLeman
Sandra I. Musanti
Barbara Trujillo
Leslie H. Kahn

C OLLABORATIONS ARE an important facet of institutional life, as they achieve outcomes that cannot be reached individually (Huxham and Vangen 2000). Successful professional collaborations build on the development of trust, the view that all participants are equal members, and the recognition of each participant's strengths and resources (Nelson and Slavit 2007). When researchers and teachers view each other as bringing valued knowledge to the process, collaborations increase the learning of all participants. Collaborations also provide opportunities for individuals to become more active within various areas of the educational community. For example, teachers can share their knowledge (e.g., at workshops or conferences) or researchers can witness what happens in a classroom. Finally, collaborations can help break down the division of traditional roles held in research settings (Raymond and Leinenbach 2000), namely, that teaching and conducting research are mutually exclusive events.

Examples of professional development (PD) experiences that develop around collaborative efforts between university faculty and K–12 teachers include areas such as lesson planning (Dumitrascu and Horak 2008), methods of assessment (Clark et al. 1996), and curriculum-wide adoptions (Arbaugh 2003). Research has also shown that Teacher Study Groups (TSGs) hold the potential to foster learning communities that create a context for collaboration (Crespo 2006). As part of the Center for Mathematics Education for Latinos/as (CEMELA), a three-phase, cross-site task adaptation and implementation occurred in spring 2007 between two different TSGs at two different sites. This chapter shares what was learned as a result of this professional collaboration, in particular focusing on the teachers' reflections and interactions.

The Teacher Study Groups

Similar to Arbaugh's (2003) notion, each TSG in this collaboration consisted of a "group of educators who [came] together on a regular basis to support each other as they work[ed] collaboratively to both develop professionally and to change their practice" (p. 141). This group of educators included CEMELA researchers (e.g., graduate and postdoctoral fellows, teacher leaders, and principal investigators) and elementary teachers of Latina/o students. For the TSGs discussed here, intentional reflection on practice through a lens of improvement (Schön 1983) was used as the forum in which to foster these communities. For both the teachers and researchers involved in the TSGs, the desire was to learn about the mathematical understandings of Latina/o students.

The composition of each study group is outlined in table 6.1. The schools from both sites served predominantly Latina/o students in low socioeconomic communities with a high percentage of English language learners (ELLs). (In this context, we refer to English language learners as those students who are still developing a proficiency in English and speak a language other than English at home.)

Table 6.1

Distribution of the participants from each TSG

	TSG1 (New Mexico)	TSG2 (Arizona)
No. of schools	3	3
No. of teachers by gender:		
Female	6	10
Male	2	0
No. of teachers by ethnicity:		
Latina/o	5	7
Caucasian	3	3
Grades	K–5	3–6
Years of experience	From 5 to 20	From 3 to 27
No. of researchers	3	6

The TSGs constituted an ideal format for partnership. On one hand, teachers furthered the researchers' understanding of instructing Latina/o students by providing access to their classrooms, sharing their expertise regarding instructional decisions made there, and providing samples of student work. On the other hand, researchers in their role of facilitator offered a PD opportunity to teachers, supporting them in areas of mathematical knowledge and assessment of students' work.

At the time of the cross-site experience, both TSGs had been functioning for over a year, with relatively little fluctuation in membership. The stability of the PD experience, built upon established norms of rapport and trust, allowed teachers to experience a sense of collegiality. This collegiality showed itself in subtle cues, such as the use of the first person plural (i.e., *we*) when discussing each others' collective work within the TSG or when referring to the presence of a researcher during classroom experiences. As another example, a visit by two researchers from TSG2 to the phase-three session at TSG1 did not seem to inhibit teachers' participation. The teachers appeared to share without reservations their recollection of the events in their classrooms and their reasons for the instructional decisions they made. However, during the phase-one session at TSG2 there was a visitor who was not a member of either TSG. This visitor worked with a group of teachers, including a normally vocal teacher. During this session, though, this teacher did not participate as much as usual. We are uncertain about what could have accounted for the teacher's reservation, whether it was her unfamiliarity with the visitor, a resistance to participate in the project, or something else unrelated to the PD experience.

The Three-Phase Experience

Sharing design elements with the SATRR model (Crespo and Featherstone 2006), a three-phase experience was created for this cross-site initiative. During the first phase, teachers explored a task on area and discussed various solutions with each other. During phase two, all of the teachers implemented the task in their classroom. Finally, in phase three, teachers examined various aspects of their instructional practice and co-analyzed student work as a result of the implementation of the task.

The task utilized was a fourth-grade National Assessment of Educational Progress (NAEP 1996) geometry measurement problem specifically asking students to compare the areas of two figures (fig. 6.1) using pictures and words to explain their solution. Current research suggests that Latina/o students can participate meaningfully in rich mathematical tasks through their mathematical discourse (Anhalt, Fernandes, and Civil 2007). We use "mathematical discourse" here as explained in Moschkovich (2002), where learning mathematics is described as participating in a community of practice; developing classroom sociomathematical norms; and using multiple material, linguistic, and social resources to solve the task.

The comparison of areas problem is a rich mathematical task that can provide multiple opportunities for teachers to reflect on various components of their students' knowledge as well as what pedagogical moves they make in their instruction of the task.

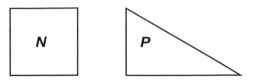

Bob, Carmen, and Tyler were comparing the areas of N and P. They each conclude the following:

Bob: N and P have the same area;
Carmen: The area of N is larger;
Tyler: The area of P is larger.

Who was correct? Use pictures and words to explain why. (Cutouts of N and P are given with the base of P being twice the side of the square.)

Fig. 6.1. The comparison of areas task

Phase 1: Teachers Experiencing the Task

Before considering how to adapt the task to meet the needs of their students, all of the teachers engaged with the task as mathematical learners. Teachers worked on the task in small groups, discussing with each other various aspects of their solution process. For example, one teacher at TSG2 conferred with other participants regarding what word to use to refer to the longest side of a right triangle (hypotenuse). It was this type of mathematical language, she shared, that she expected of her students, which is why she wanted to be as precise as possible in her writing.

The teachers then reflected both in small groups and in whole-group discussions on how their Latina/o students might experience the task. During these discussions, the teachers considered the mathematical knowledge needed by the students to solve the task. All of the teachers acknowledged that their students would need multiple and varied experiences with the concept of area to solve the task successfully. Some of the teachers indicated that even if their students possessed a solid understanding of area, some might be challenged by the inexactness of the cutout shapes given to them. (The manipulatives used in the task implementation were paper triangles and squares cut by CEMELA researchers, and thus a number of shapes had inexact measurements.) However, all teachers felt that providing their students with concrete manipulatives was an important consideration for their classroom practice. As had been discussed in prior TSG sessions at both sites, and as research supports (e.g., Bay-Williams and Herrera 2007; Moschkovich 2002), providing all students, but especially ELLs, manipulatives increases students' ability to develop a mathematical discourse around specific concepts. This is because manipulatives are both concrete objects on which to develop the concepts and also tools through which students can communicate about those concepts.

Across sites, teachers also considered the potential linguistic demands for their students. From her experience of solving the task, one TSG2 teacher, Ms. Castillo, commented that the directions indicated to her that she could only use one of each shape to arrive at the solution. Because of these directions, she felt that "kids [wouldn't] take more shapes" and thus would not think to solve the problem in some of the ways seen in the TSG. In response to this, another teacher, Ms. Domínguez, commented that she too had been very literal in her interpretation of another problem. In that case, Ms. Domínguez was given the directions to find the area of the shaded region of a particular picture. However, to find the area, Ms. Domínguez had to consider the unshaded region of the picture, something she did not do because the directions did not specify that region. Ms. Domínguez noted that in both situations she and Ms. Castillo had allowed the wording of the problems to direct their thinking into not considering other strategies beyond the one implied by the words in the directions. As a result, Ms. Domínguez wondered how to present problems so that the language used would not confine students' thinking.

It is important to think about how the wording of directions may influence students' mathematical thinking. Moreover, we must consider how much direction to provide students. Ideally we want to provide students with directions in which they are not told how to solve a problem but rather are challenged mathematically. In reality, though, finding the specific words to do that is difficult. For example, if Ms. Domínguez had been told to consider the unshaded region in the directions, the task might have been stripped of its problematic nature. This issue is further intensified when considering ELL students. Teachers may unnecessarily complicate the task by either giving vague or restrictive directions that can obstruct the students' learning of mathematics.

Ms. Domínguez's insight was achieved as a result of a collaborative environment with teachers working together to share their thoughts and expertise. Before this study group session, Ms. Domínguez had often looked for the best way to present problems to her students, but it appears she had not considered how her students' mathematical thinking might be confined by the specific directions in a problem until Ms. Castillo shared her experience and opinion.

Some teachers noted that their students would probably encounter difficulty when justifying their thinking in words. Other teachers focused on the need to build on their students' knowledge of mathematical vocabulary in Spanish by using cognates from English to Spanish (words in both languages that have a similar spelling and have the same meaning such as perimeter/*perímetro* or rectangle/*rectángulo*). The teachers shared that issues may arise owing to the use of language from different registers, in particular everyday language and the language specific to mathematics. A register is a variety of a language (words, phrases, modes of communication, etc.) used for a particular purpose or in a particular social setting or context. For example, a *mathematics register* would refer to meanings of words that are used in mathematical situations, such as a mathematics classroom (Halliday 1978). Ms. Saavedra from TSG2 hypothesized that her students might struggle to make sense of the mathematical notion of area (as the number of square units in the interior of a closed figure) as a result of their everyday experiences with the word *area*, as when they say, "This area is mine."

Another conversation at TSG2 centered on specific confusions that Spanish-speaking students might encounter, in particular how to say "flip," "rotate," and "turn" in Spanish. Based on these conversations, it appears that the teachers were not oversimplifying the task of educating diverse learners (Rothstein-Fisch, Trumbull, and Greenfield 1999). Instead, they addressed their students' mathematical and linguistic development (Bay-Williams and Herrera 2007; Moschkovich 1999) by discussing how to promote student participation in mathematical discourse through the use of their native language and concrete artifacts like manipulatives. The goal was to enable students to develop concepts, negotiate meanings, and engage in meaningful mathematical conversations.

Phase 2: Teachers Adapting and Implementing the Task

After investigating a particular task, the teachers collectively considered how to adapt and implement the problem within their own classroom. Here they discussed logistical concerns (such as how many shapes should be provided to their students) along with what tools should be made available during implementation. For example, reflecting on their experiences and conversations regarding the inexactness of the cutout shapes, some teachers in TSG1 considered using the word "quadrangle" instead of "square." All decisions were discussed among the participants. Researchers posed questions to stimulate conversations about such considerations as the role of language, prior knowledge, which tools would be allowed, and so on. Teachers shared their specific concerns, and some agreed to follow similar guidelines, such as giving the same number of pieces to the students and forming groups. Other conversations concerned how to adapt the task to different grade levels. For example, kindergartners would identify shapes and compare similar (same shape, different size) objects to determine which one is bigger, while building on their skills for explaining their answers and listening. Third-grade teachers would need to introduce the concept of area itself at this point, while fourth graders would need only to undergo a review of the concept of area before being given the task.

During the week between study group sessions, each teacher implemented the task in his or her classroom, with both teachers and researchers participating in the classroom activities. In most cases, the teachers implemented the task as researchers either observed or videotaped the implementation. One teacher from TSG2 asked the researcher to introduce the task, stating that she needed to take care of administrative tasks. In the spirit of collaboration, the researcher resolved this unexpected situation by introducing the task.

In implementing the task, all of the teachers utilized group work. While the students worked on the task, they were free to get whatever materials and manipulatives they needed. Some teachers were more deliberate than others in making the tools available (putting them on the tables, for example, instead of having the students go to the cabinets to retrieve them), as well as in the number of squares and triangles given to individual students. After the students finished working on the task, the students provided solutions and justified their thinking to the entire class, a standard practice for these teachers.

Phase 3: Teachers Reflecting on the Classroom Experience

The third phase allowed teachers and researchers to co-analyze the instructional decisions made during the task implementation as well as the work produced by the students. Collectively, the teachers shared how they introduced and clarified the task to support students' construction of knowledge. Among the elements discussed were how teachers encouraged students to work with the group at their table and across groups in the classroom to come to a shared understanding; made specific adaptations to help ELLs to gain entry into the task; and required their students to prove their conjectures regarding their solution of the task. As opposed to simply explaining the solution, Mr. Sloan (TSG1) shared: "[Prove] means that they [students] have to, either in writing or using a manipulative or orally, demonstrate that their answer is correct. Because *explain* is kind of fuzzy. [They can say that] the rectangle is bigger because it looks bigger. And they explained the answer. So I use the word *prove* so that they know there has to be a clear demonstration that can't be argued."

In conjunction with the emphasis on higher cognitive demands (e.g., problem solving, hypothesizing, proving), the teachers made specific instructional decisions based on their knowledge of the students in their classrooms. Some teachers adapted their introduction (e.g., reading the problem aloud and providing visual cues) to support their ELLs' access to the task. Other teachers reviewed mathematics vocabulary, used students' vocabulary (including Spanish), and rephrased students' answers. For example, in Ms. Green's (TSG2) sixth-grade class, the students had been using rectangular arrays to study factors of numbers. When a student in her class was confused about how to calculate the area of a square, Ms. Green referenced this prior mathematical concept and terminology by guiding the student to consider the areas of different arrays.

Implementing the task and participating in the collaborative discussion led a couple of teachers to consider modifying specific elements of their teaching. At TSG1, one teacher considered expanding the dialogue that took place in the study group session by showing his students the videotaped task implementation of his classroom. At TSG2, after hearing how other teachers provided manipulatives to their students, one teacher reflected on how her manner of providing manipulatives might have impacted her students' successes or challenges with the task. Instead of placing a pile of shapes in the middle of the table where some students took the shapes and some did not, the teacher wondered if giving each student a set number of shapes each might prompt more students to use the manipulatives.

Researchers and teachers also reflected on various aspects of students' thinking through the co-analysis of student work. During this process, the teachers confirmed many of the expectations they shared during the first session, as well as noticed some surprises regarding their students' thinking. An example of the realization of teachers' expectations is the student work depicted in figure 6.2. It is possible that the unequal length of the shorter side of the triangle (P) and the side of the square (N), a consequence of the in-exact cutout shapes, led the student to limit her thinking to the comparison of the lengths of the sides only. As another example, Ms. Saavedra noted that some of her students' work indicated that they did need more experiences with the concept of area, as they confused area and perimeter, a follow-up to the conversation held during the phase one session at TSG2. Teachers in TSG1 came to the identical conclusion.

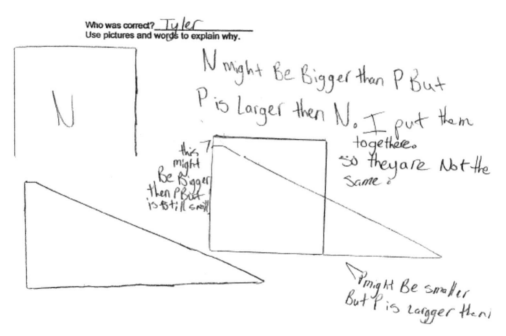

Fig. 6.2. Sample of student work

Teachers in both TSG1 and TSG2 noted the persistence of their students to solve the task. For one teacher at TSG1, Ms. Kim, this task showed her how diligently and intensely her students work, a characteristic appreciated by another teacher in the study group: "It seems that one of the things I hear teachers being frustrated with is that they give a problem and automatically [the students say] 'I don't get it' instead of trying to figure it out." It seems that the NAEP task allowed students a sense of agency because there were multiple entry points into the problem, depending on the student's level of mathematical ability, but at the same time, it allowed for multiple solution strategies.

In Closing

As participants in this PD experience, teachers adapted their instruction to make a shared NAEP task challenging and accessible to their Latina/o students. They discussed their expectations for and surprises regarding their students' work. They reflected on the linguistic demands of mathematics instruction, such as navigating different registers or the problems posed by homophones (words that are pronounced the same way but have different meanings, like *two* and *too)*. They created contexts for higher-order thinking and oral and written communication, which supported students in their learning and development of reasoning (Khisty 1997).

The investigative experience described in this chapter showcases several elements of a successful professional collaboration described in the literature (Arbaugh 2003; Crespo 2006; Musanti and Pence 2010; Nelson and Slavit 2007). It took place in the context of a sustained professional development model among a community of practitioners. The focused, ongoing collaboration generated spaces where teachers could examine linguistic and mathematical issues related to instruction, and all members could learn through reflection and dialogue with colleagues. Finally, it centered on the teachers' practice, which provided them an experience that was situated and relevant.

Collaborations like this that occur over time allow for the development of a mutual respect for each others' expertise and increase the learning of all participants. Researchers and teachers view each other as colleagues that bring valued knowledge to the process. Under these conditions, researchers and teachers engage as a community through shared conversations centered on student work and instructional practices. By having a common experience such as investigating a mathematical task, teachers can jointly reflect on the demands facing students and explore opportunities to provide students. Researchers have the opportunity to learn from teachers about the issues with which they grapple as they plan and enact their instruction. For both populations, collaborations are a valuable tool which can further the collective understanding of effective mathematics instruction.

······

We would like to thank Marta Civil and Rick Kitchen for their help in making the study groups possible and in creating their design, and for their advice, revisions, and encouragement throughout the writing of this chapter.

CEMELA is a Center for Learning and Teaching supported by the National Science Foundation, grant number ESI-0424983. Any opinions, findings, and conclusions or recommendations expressed in this chapter are those of the authors and do not necessarily reflect the views of the National Science Foundation.

REFERENCES

Anhalt, Cynthia, Anthony Fernandes, and Marta Civil. "Exploring Latino Students' Understanding of Measurement on NAEP Items." In *Proceedings of the Twenty-Ninth Annual Meeting of the North American Chapter of the International Group for the Psychology of Mathematics Education*, Vol. 3, edited by Teruni Lamberg and Lynda R. Wiest, pp. 677–80. Stateline (Lake Tahoe): University of Nevada, Reno, 2007.

Arbaugh, Fran. "Study Groups as a Form of Professional Development for Secondary Mathematics Teachers." *Journal of Mathematics Teacher Education* 6 (2003): 139–63.

Bay-Williams, Jennifer M., and Socorro Herrera. "Is 'Just Good Teaching' Enough to Support the Learning of English Language Learners? Insights from Sociocultural Learning Theory." In *The Learning of Mathematics,* 2007 Yearbook of the National Council of Teachers of Mathematics (NCTM), edited by W. Gary Martin and Marilyn E. Strutchens, pp. 43–63. Reston, Va.: NCTM, 2007.

Clark, Caroline, Pamela A. Moss, Susan Goering, Roberta J. Herter, Bertha Lamar, Doug Leonard, Sarah Robbins, et al. "Collaboration as Dialogue: Teachers and Researchers Engaged in Conversation and Professional Development." *American Educational Research Journal* 33, no. 1 (Spring 1996): 193–231.

Crespo, Sandra. "Elementary Teacher Talk in Mathematics Study Groups." *Educational Studies in Mathematics* 63 (2006): 29–56.

Crespo, Sandra, and Heather Featherstone. "Teacher Learning in Mathematics Teacher Groups: One Math Problem at a Time." *AMTE Monograph, vol. 3: The Work of Mathematics Teacher Educators* (2006): 97–115.

Dumitrascu, Gabriela, and Virginia Horak. "Lesson Study: A Site for Teachers' Professional Growth and Use of Instructional Innovations in ELL Classrooms." In *Proceedings of the California Capital Lesson Study Conference,* pp. 55–75. Sacramento: California State University, Sacramento, 2008.

Halliday, Michael Alexander Kirkwood. "Sociolinguistics Aspects of Mathematical Education." In *Language as Social Semiotic: The Social Interpretation of Language and Meaning,* edited by Michael Alexander Kirkwood Halliday, pp. 194–204. London: University Park, 1978.

Huxham, Chris, and Siv Vangen. "Ambiguity, Complexity, and Dynamics in the Membership of Collaboration." *Human Relations* 53, no. 6 (2000): 771–806.

Khisty, Lena L. "Making Mathematics Accessible to Latino Students: Rethinking Instructional Practice." In *Multicultural and Gender Equity in the Mathematics Classroom: The Gift of Diversity,* 1997 Yearbook of the National Council of Teachers of Mathematics (NCTM), edited by Janet Trentacosta and Margaret J. Kenney, pp. 92–101. Reston, Va.: NCTM, 1997.

Moschkovich, Judit N. "Supporting the Participation of English Language Learners in Mathematical Discussions." *For the Learning of Mathematics* 19 (1999): 11–19.

————. "A Situated and Sociocultural Perspective on Bilingual Mathematics Learners." *Mathematical Thinking and Learning* 4, nos. 2 & 3 (2002): 189–212.

Musanti, Sandra I., and Lucretia (Penny) Pence. "Collaboration and Teacher Development: Unpacking Resistance, Constructing Knowledge, and Navigating Identities." *Teacher Education Quarterly* 37, no. 1 (Winter 2010): 73–89.

Nelson, Tamara H., and David Slavit. "Collaborative Inquiry among Science and Mathematics Teachers in the USA: Professional Learning Experiences through Cross-Grade, Cross-Discipline Dialogue." *Professional Development in Education* 33 (2007): 23–39.

Raymond, Anne M., and Marylin Leinenbach. "Collaborative Action Research on the Learning and Teaching of Algebra: A Story of One Mathematics Teacher's Development." *Educational Studies in Mathematics* 41 (2000): 283–307.

Rothstein-Fisch, Carrie, Elise Trumbull, and Patricia Greenfield. *Supporting Teachers to Bridge Cultures for Immigrant Latino Students.* San Francisco: WestEd, 1999.

Schön, Donald A. *The Reflective Practitioner: How Professionals Think in Action.* New York: Basic Books, 1983.

CHAPTER 7

The School-Based Teacher Learners-to-Leaders Project

Learning from an Authentic Problem of Practice

Jon Rahn Manon
Janice McCarthy

Wow! What an absolutely rich and fulfilling experience. I feel that I got more out of one single SBTL workshop than I got out of all my other professional development days combined in my five years of teaching. This program, and the people involved, has had such a profound impact on my teaching practice that words cannot express how much the entire experience has meant to me. Everything from breaking down video, to doing the math, to district meetings, to regional meetings, to state-wide meetings, to SBTLfests (past and present) have been beneficial to me.

W HILE THESE WORDS, penned in a final reflection by a teacher participant in our School-Based Teacher Leaders (SBTL) project, may seem like extravagant praise, they do speak to this particular teacher's need for professional community and to the ways in which our project fulfilled that need. In this chapter we will examine the features of the SBTL project that provided this teacher and many of his colleagues with opportunities for professional learning and growth. We will also describe how this community was created, how its resilience was challenged, and, ultimately, why it remained vibrant over a five-year period. The SBTL project was funded by a series of state Mathematics and Science Partnership (MSP) grants awarded to our math and science resource center at the University of Delaware in partnership with a dozen local school districts. It enrolled nearly 140 middle and high school teachers of mathematics over its lifespan and, most important, produced positive outcomes for many students at risk for failure in secondary mathematics. This is the story of a professional development effort that grew out of an authentic problem of practice and that was fueled, in turn, by our teachers' willingness to make their practice public through the very powerful medium of classroom video.

Professional Development Using Real Classroom Contexts

One persuasive theory of teacher learning posits that professional development will be most effective if situated in authentic activities derived from actual classroom contexts (Putnam and Borko 2000). This view also holds that cognition is social and, therefore, distributed across the members of a learning community and that these factors should be considered when structuring professional learning opportunities. To facilitate teacher learning, it has been suggested that a curriculum and pedagogy for professional development are needed (Ball and Cohen 1999). This curriculum should be centered in inquiry into practice, provide comparative perspectives on practice, and contribute to professional inquiry. Thompson and Zeuli (1999) and Ball and Cohen (1999) propose similar components for a pedagogy of professional development. These include: a community that is willing to learn together; worthwhile tasks that create cognitive dissonance; tools for studying teaching practice; a focus on inquiry into student learning; the development of new practices that align with the new understandings about student learning; support in enacting changes; and an understanding that learning to teach is a lifelong journey.

It is on these assumptions about a generative environment for professional learning that we predicated our five-year-long professional development project in secondary mathematics. We named the project School-Based Teacher Leaders, with the optimistic premise that teacher participants might become exemplars of reflective practice in their schools and districts. A unifying belief of our leadership team, made up of the authors as well as other mathematics educators working for our math and science resource center, was that a program of professional development should respond directly to an authentic and pressing problem of practice.

An Authentic and Pressing Problem of Practice

A central assumption underpinning our design for professional development is that teacher learning is most effective when it is problem-based. The University of Delaware's Mathematics & Science Education Resource Center (MSERC) was begun in October 1996 in order to "support the development of standards-based instruction in mathematics and science in Delaware's public schools." The educators in our center have long advocated for the problem-based learning of mathematics; we believe that students can achieve a deep and durable understanding of mathematics through wrestling with and solving challenging mathematics problems (Erickson 1999). We have also come to believe, through our experience with the School-Based Teacher Leaders project, that teachers can become more effective at their craft by struggling with and solving challenging *problems of professional practice*. This analogy also points to another important feature of a problem-based pedagogy: Just as students of mathematics benefit from working collaboratively with peers at certain junctures during the problem-solving process, so will teachers benefit by working with colleagues in a supportive community as they tackle particularly obdurate problems of pedagogical practice (Ball and Cohen 1999).

In fall 2003, the Delaware secretary of education convened a group of educators, including this chapter's first author, to study the challenges of middle-grades mathematics in our state. A report issued the next spring entitled *A Blueprint for Action* concluded that "a belief change is needed, replacing notions that some students simply can't learn mathematics with a far more empowering belief that all students can master critical mathematics concepts." According to the focus group's report, "Improvement in mathematics achievement must start with more reflective teacher practice, based on detailed reviews of collected student work, collaboration with colleagues on lesson improvement, and content-rich training in mathematics instruction."

The focus group's report, in turn, was used as the basis for our state department of education's first call for Mathematics and Science Partnership proposals. After several planning meetings with district curriculum specialists, our center submitted a proposal in partnership with twelve of our state's seventeen school districts. We chose to build our professional learning community around the problem of students at risk for failure in middle-grades mathematics. Despite the implementation of NSF-funded exemplary curriculum materials in many middle schools in our state, far too many of our students were still failing to learn mathematics in meaningful ways. Student achievement on the Delaware Student Testing Program's math exam had reached a plateau at around 60 percent meeting or exceeding the standard. It was the 40 percent not meeting the standard that became the central problem of practice for the teachers in our project.

Nested Learning Communities

Our project involved successive cycles of action research conducted by our corps of secondary math teachers who had volunteered or been nominated to serve as school-based teacher leaders. Our professional development model was built on a structure of nested professional learning communities. At the school level, teachers met individually with their district math specialist, who entered the classroom as observer and coach and helped with the videotaping of teacher-selected mathematics lessons. At the next level, teachers met in study groups within their district, facilitated by their math specialist. These district teams were then further "nested" within evening regional meetings, and the entire project came together in all-day statewide forums several times each year, including the culminating "SBTLfest" in late spring. SBTLfest, as it came to be called, was a mini-conference held each spring at which district teams presented the results of the year's action research. Keynote speakers over the years included James Hiebert, Diane Briars, Alfinio Flores, and Cathy Seeley.

Perhaps the most daunting challenge for our project at its inception was how to organize the research across schools, teachers, and at-risk students. We wanted to understand the factors that put students at risk for failure in secondary mathematics, and we also wanted to have our participating teachers pilot instructional interventions to address these risk factors. Therefore, we chose an "action research" approach. Action research presumes that the practitioner-researcher will modify her or his professional practice in response to the evidence collected in the enactment of that practice (Mills 2000). Our working definition of "at-risk students" was essentially empirical. By and large, students scoring below standard on our state assessment were considered at risk for failure in secondary mathematics. The fact that nearly 40 percent of eighth-grade students, for example, scored at Performance Level 2 ("below standard") or Performance Level 1 ("well below standard") on the mathematics component of our state test suggested the magnitude of the challenge. Clearly, we could not study *all* of the at-risk students in every one of the fifty-plus project teachers' classrooms.

We realized that our participating teachers needed to be personally invested in the project. To achieve this, we decided to have each teacher select two or three focus students from among those considered at-risk in one of their classrooms. This might limit the breadth of the inquiry, but it would add a much-needed depth. We framed this project as composed of dozens of action research projects, each located in the classroom of a participating teacher. While understanding the results of the myriad action research efforts proved to be a real challenge, the decision to have each teacher make a principled selection of one or more focus students proved central to the ultimate success of the project.

Classroom Video: A Tool for Studying the Practices of Teaching

Without a doubt, our choice of classroom videotape as the best tool for studying local examples of teaching and learning with at-risk students gets much of the credit for the vitality, longevity, and productivity of our project. Teacher testimony makes this very clear. According to one sixth-grade teacher who remained in the project for several years, "We turn the cameras on our at-risk students but, at the end of the day, this project is really meant to help us think about our teaching and become better teachers." Another teacher added, "Videotaping allowed me to see some really great things happening in small-group interaction that I frequently did not realize was occurring. It gave me the *evidence* I need to continuously change and improve my practice."

As these comments suggest, the protocol for using video in this project was that the camera was to be trained on the focus students rather than on the teacher. In an earlier study we conducted using classroom video, the camera tracked the teacher . This time, our intent within the SBTL action research was to record student behavior and utterances; we then used these to try to make inferences about student thinking both when the teacher was present and, even more revealingly, when he or she was out of the frame. As it turned out, our teachers learned a great deal about the until-now secret world of student action and thinking in their classrooms through this tactical approach to data collection.

Risk Factors Observed

In many cases, SBTL teachers selected focus students based on past records of failure in mathematics. In other instances, they paid special attention to behaviors observed during the first weeks of the new school year. Students who seemed to exhibit diminished or negative affect, little or no engagement in mathematical tasks, or conspicuously off-task behavior were often chosen for closer scrutiny. Teachers who were in the project over multiple years frequently found themselves selecting for different risk factors from one year to the next as they sought to understand a greater range of the factors inhibiting success in mathematics in their classrooms.

For our part, the project leadership reminded the participants of the need to collect evidence to support their conclusions about student risk factors and possible instructional responses. This method was modeled at each regional or statewide professional development event, and careful observation of at-risk student behaviors quickly became the norm.

Through observation and video analysis, teachers catalogued a variety of behaviors of at-risk students that interfered with their problem solving and ultimately compromised their ability to learn mathematics. In most cases, these behaviors were observed across students and classrooms and included both avoidance and dependence behaviors. Avoidance behaviors included not only attempts at actual flight from the problem-solving setting (such as requests to leave the room to go to the bathroom), but also another class of behaviors that we have described as "attempts to appear on task." These included elaborate preparations before beginning the day's assignment, frequent trips to the pencil sharpener, a ritualized organization and reorganization of materials, and repetition of nonmathematical tasks such as repeatedly drawing and erasing a Cartesian coordinate system in apparent preparation for plotting points.

Dependence behaviors were also exhibited by many of our at-risk students; these were characterized by reluctance to attempt the mathematics without explicit direction from the teacher. At-risk students exhibiting these behaviors tended to avoid interactions with peers, even during times designated for small group collaborative work. Such students were easily observed as they often had their hands raised in an attempt to attract the teacher to provide direction and help.

Identifying and Implementing Instructional Interventions

In order to promote greater engagement and, ultimately, greater success on the part of their identified at-risk students, our school-based teacher leaders piloted some instructional interventions intended to address the behaviors that interfered with learning. A range of interventions were piloted and catalogued, but most of these shared a number of common attributes.

Most interventions emphasized teacher actions that (a) communicated higher expectations to focus students, (b) encouraged at-risk students to work productively with their peers, or (c) reduced or eliminated the nonmathematical elements of classroom tasks. Potential interventions were discussed at monthly regional or statewide meetings and illustrated with video vignettes whenever possible. Many of our teachers' action research projects ultimately coalesced around a few common foci. The most frequently cited of these went to the larger question of how to help at-risk students move from dependent to interdependent and independent mathematical problem solving. As videotapes of small-group problem solving were closely observed and analyzed, our project teachers identified conditions under which productive interdependence seemed to emerge. They looked at their own role in this process, the centrality of a truly challenging task, and other factors that seemed to impact small-group dynamics.

Upon viewing a videotape of their own classrooms, teachers frequently commented that when they approached a group of students working on a math problem, they witnessed themselves going almost immediately into strong scaffolding mode. Typically this was before they had accurately assessed the state of progress of the group itself. Teachers would often, they noticed, single out an at-risk student from the group and provide one-to-one instruction meant to move that student toward successful completion of the task at hand. In other words, the teachers in the project came to believe that their interactions were often detrimental to productive group dynamics, and they sought strategies for modifying their approach to small-group interventions. One simply stated strategy that became very popular in our project was the "ask three before you ask me" response to student questions. This promoted interdependence within the group itself and explicitly reminded students that peer-to-peer collaboration, rather than teacher answer-giving, was the new classroom norm.

At our January statewide meeting during the second year of our project, our SBTL teachers were tasked with developing a small-group discourse rubric (fig. 7.1). This rubric defined norms for small-group work; it was ultimately translated into several student-friendly versions that were used by many teachers to help structure expectations for their students around productive collaborative behaviors.

Teacher Learning

We found compelling evidence that some of our school-based teacher leaders did in fact adopt "new practices" after engaging in this work over a number of years. Self-reports were routinely encouraging. According to one teacher participant: "I can't imagine where I'd be professionally without having been a part of the SBTL group. Being involved with SBTL has changed my teaching and my view of others' teaching. When we joined STBL, none of us knew what we were getting into; we didn't know how much it would impact our craft."

SBTL Student-to-Student Small-Group Discourse Rubric		
	Quality of Student-to-Student Discourse	**Mathematical Content**
4	All students are engaged in talking about the reasoning and problem-solving process. Students question each other's ideas and encourage each other to justify their reasoning. Together they reflect on multiple ways of looking at the problem. Talk about group norms may be a part of the process, but this is done efficiently and does not interfere with mathematical thinking.	All students successfully complete the mathematical task and are able to explain/defend their group's solution(s). Although there may be different strategies used within the group, students are knowledgeable about each other's reasoning.
3	Students talk in meaningful ways about the mathematics. This may be interspersed with students working at their own pace. The group may occasionally become sidetracked by the need to define roles and encourage participation by all.	All students are able to complete the mathematical task. However, some are better than others at explaining or defending the reasoning and may not be able to justify the group's decision.
2	Minimal collaboration between two or more students or superficial group and individual interactions. Group as a whole is on task, but some students are not contributing. Students do not attempt to bring each other to an understanding of the mathematical task or the collaboration norms.	One or more students do not succeed on the mathematical task. The group as a whole has a fragmented understanding of the task, and one or more students is unaware of the others' strategies. There is a correct answer to the task, but it may have been done by only part of the group, then mimicked by others.
1	In general the group is working on the task, but they are not communicating or working as a group. This could be characterized by silence or by bickering about group roles. This includes groups that ask each other, "What did you get?" or simply check each other's solutions.	Students are not successful in completing the task collaboratively. The solution is incomplete, incorrect, or not justified by the majority of students within the group.
0	There is little or no dialogue about the understanding of the mathematical task. Students become completely off-task or disassociated from the problem at hand.	Students do not address the mathematical task as a group.

Fig. 7.1. Teacher-developed small-group discourse rubric

In what ways did entering into the SBTL project change teachers' practice and "impact their craft?" According to teacher self-reflections, the SBTL teachers tended to reconceptualize the abilities of their students. One teacher, who had participated in our project for the full five years, said the following about how his teaching had changed: "One of the most important things I learned from my experience in SBTL is that, too frequently, I allowed my expectations for students to be tempered by their own expectations for themselves." And, then, later: "There was clear growth in my practice of teaching math over the past five years. I increasingly focused on student learning as guide to revising my teaching. . . . There was continued movement toward cooperative and collaborative learning as the classroom became more student focused and less teacher focused."

Another teacher used the rather dramatic metaphor about "rescuing his students from a burning building." He said he no longer believed that he could ignore the needs of *any* of his students—he had to "get them all out."

Other evidence corroborates these changes in teacher practice. Using a classroom observation protocol called Determining the Quality of Mathematics Instruction that was developed in collaboration with our evaluator (Stazesky 2004), we collected data on teacher practice. This protocol included eighteen indicators of an effective problem-based pedagogy across three broad domains: the Design and Implementation of the Lesson (eight indicators), Mathematics Content (five indicators), and Classroom Culture (five indicators). Descriptors within each of the indicators were designed so that a trained observer could rate the indicator as "close to ideal," "getting there," or "not even close." The scores pertaining to each indicator were then aggregated across all observed teachers within a given observation cycle, and the percent of teachers (not lessons) scoring at each of these levels was calculated and represented graphically (fig. 7.2).

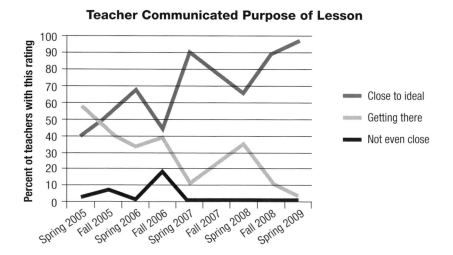

Fig. 7.2. Change over time in a key instructional indicator

A powerful legacy of our project is that this instrument was utilized across the five years of grant funding—once in the spring of 2005 to establish baseline data, and then twice for each of the four succeeding years for a total of 336 discrete classroom observations. These repeated measures allowed us to look for growth within each year and change across the years, and we found substantial progress on several key indicators. For example, for the first indicator in the Design and Implementation of the Lesson domain ("Teacher clearly defines and communicates a purpose for the lesson"), only 40 percent of teachers were "close to ideal" in the initial observation, but 90 percent were "close to ideal" by spring 2007, as were nearly 100 percent of teachers in the final observation taken in spring 2009.

Table 7.1 compares spring 2005 baseline measures for each level of each indicator to values attained in the final set of observations recorded in spring 2009. Other indicators showing substantial growth included (1) *Teacher plans and/or adjusts instruction based on students' level of understanding;* (2) *The content is balanced between conceptual understanding and procedural fluency;* and (3) *Active participation of all is expected and valued* (Uribe-Zarain 2009). Two indicators in which we found the growth to be less satisfying were (1) *The teacher advances the development of student understanding;* and (2) *Elements of mathematical abstraction are included when appropriate to do so.* Particularly concerning was the fact that, on average, little or no growth was seen across the life of the project on a third indicator that monitored whether *Intellectual rigor and/or the constructive challenge of ideas are evident.* We hypothesize that, especially when working with at-risk students, teachers may incrementally reduce the cognitive demands of the tasks given to students for problem-based learning (Stein et al. 2009).

Table 7.1

Longitudinal change in observational protocol

Observational Protocol Indicators: Baseline (2005) to Project Conclusion (2009)						
Indicators ⟍ Incidence	Close to ideal		Getting there		Not close	
	2005	2009	2005	2009	2005	2009
D1. Teacher defines and communicates purpose	40%	96%	58%	4%	2%	0%
D2. Teacher engages students with important ideas	36%	63%	31%	25%	33%	12%
D3. Adequate time and structure for investigation	25%	53%	15%	10%	60%	37%
D4. Adequate time and structure for "wrap-up"	26%	34%	10%	48%	64%	18%
D5. Achieves a collaborative approach to learning	30%	34%	28%	43%	42%	23%
D6. Enhances development of student understanding	20%	25%	25%	43%	55%	32%

Table 7.1—*Continued*

Observational Protocol Indicators: Baseline (2005) to Project Conclusion (2009)						
Indicators Incidence	Close to ideal		Getting there		Not close	
	2005	2009	2005	2009	2005	2009
D7. Assesses the level of student understanding	50%	58%	28%	25%	22%	17%
D8. Adjusts instruction re: student understanding	38%	96%	33%	0%	29%	4%
M1. Balance between conceptual and procedural	32%	75%	25%	10%	43%	15%
M2. Content challenging and accessible	28%	70%	30%	20%	42%	10%
M3. Teacher provides content that is accurate	37%	88%	61%	10%	2%	2%
M4. Elements of abstraction included as appropriate	10%	39%	33%	28%	57%	33%
M5. Connections made to math and real world	39%	83%	21%	4%	40%	13%
C1. Active participation of all expected and valued	18%	62%	49%	35%	33%	3%
C2. Respect for students' ideas and contributions	48%	90%	30%	8%	22%	2%
C3. Management strategies enhance productivity	50%	87%	40%	10%	10%	3%
C4. The classroom climate encourages students	18%	53%	61%	35%	21%	12%
C5. Intellectual rigor/constructive challenge of ideas	16%	22%	31%	29%	53%	49%

Students No Longer at Risk

Did the changes we measured in teaching behaviors result in enhanced learning for at-risk students? There is strong evidence that they did. During the final two years of the project, we found impressive learning gains for many of our focus students. After disappointing outcomes on the mathematics component of our state test over the first three years, we found that, in the fourth year of the project, the mean gain in performance level for our focus students scoring at Performance Level 1 ("well below standard") and Performance Level 2 ("below standard") was 0.62, meaning these students improved by more than half a performance level. This stands in stark contrast to the general trend in secondary mathematics of a *net loss* in mean performance level from grade to grade.

The results from the final year of our project were even more promising. Comparing the mean performance level on the state test in spring 2008 with the mean performance level of the same cohort of focus students in spring 2009 following a year of classroom interventions, we found a gain of 0.81 performance levels, or nearly one full performance level per student. In fact, during the 2008–2009 school year, 49 percent of the lowest achieving at-risk students gained one performance level, and 17 percent gained two full levels. Nearly half of all at-risk focus students in our study moved from "below" or "well below" to meeting the standard that year.

Our state's mathematics test is only one metric of progress, albeit one of consequence. Other measures might include changes of behavior or disposition in reluctant learners. Although these are harder to quantify, both teacher testimony and video records suggest such changes for many of our focus students. A teacher who participated for the full five years of the project described changes in the mathematical identity in one of his focus students as follows: "I felt my focus students grew a lot this year, mainly from the self-confidence they gained while in my classroom . . . This was very obvious in the case of Maria [a pseudonym] who came into my classroom as a very quiet student and never volunteered to answer any questions. After several months of coaxing, setting up successful circumstances, and nurturing positive reinforcement, she began to come out of her shell and take charge of her thoughts."

The fact that we did not see consistent and impressive student gains until the fourth and fifth years of our project might suggest that something essential was missing early on and that only later, during the final two years, did we get it right. Certainly, our SBTL teachers did become more skilled at looking at and learning from classroom video, especially after we developed and implemented a modified tuning protocol for the collaborative analysis of classroom video (McDonald et al. 2007) in the final year of grant funding. But we believe that a complete explanation is more complex.

If we subscribe to a situated cognition perspective (Lave 1988) and, furthermore, with Putnam and Borko (2000), if we hypothesize that learning is social, with knowledge distributed in perhaps nonuniform ways across a community, then it makes some sense that knowledge can reside both in individual teachers and also in the community writ large. So, even though teachers joined and departed our project across the years, a fund of knowledge about at-risk learners grew within the professional learning community that was SBTL. We believe that this community-embedded wisdom of practice was generated in response to a compelling problem of practice, and that this knowledge found application more generally and by more members of that community as it was further articulated and the community matured.

Lessons Learned: Learners to Leaders?

At the outset, we chose to name our initiative the School-Based Teacher Leaders project. Our ambition was for the SBTL participants to become teacher leaders in their own schools and districts. Our apparent failure to achieve this at scale may be the biggest disappointment of this work. While teachers, even during the first and second year of the project, would dutifully use materials piloted at regional or statewide meetings with small groups of colleagues in their home school, that did little to guarantee that they would take on the mantle of teacher leader in their local context. At one point we even renamed the project SBTLL to emphasize our expectation that, over time, teachers would transition from school-based teacher *learners-to-leaders*; however, we never did find a way to promote emergent leadership in a reliable manner.

Even so, we found some bright spots in the progression of "learners to leaders." During the fourth year of our project, we collected data at two school sites to address the question of emergent leadership using social network analysis (Spillane and Diamond 2007). Teachers at one middle school and a high school in a second district were asked a series of eleven questions. In response to the prompt "Select up to three people who most often inspire you to try something new in your classroom teaching," we find the accompanying distribution of influence in the middle school studied (Ackerman and Uribe-Zarain 2008).

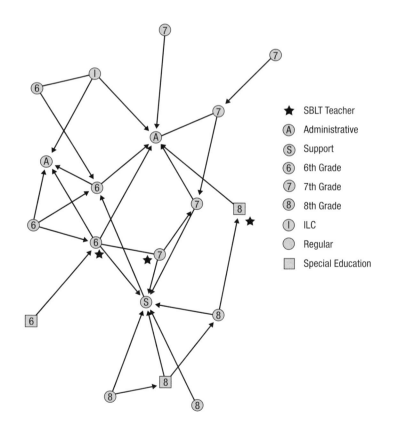

Fig. 7.3. Middle-school math social network diagram

The edges in this diagram represent connections between individuals, with the arrows indicating the direction of influence. Upon inspection of this social network diagram, we note that the SBTL participants, denoted by stars, are important "nodes" in this particular network. This would suggest that these individuals either were important "connectors" before they engaged in the SBTL project or had become influential with their peers after taking part in our professional development project. No doubt both interpretations have some merit.

Our school-based teacher leaders should, we believe, have the final word on the impact of our project. As one teacher put it, "There is no better way to learn about yourself as a teacher. It is not always easy because you cannot deny what is on the tape! So sometimes you have to face an uncomfortable truth ... but then you learn and grow."

A second teacher wrote about receiving a note from a student who said, "I will remember you challenged me in math! You thought everyone could be the best in math." This teacher goes on to say, "This was from a student who struggled every day. For her to thank me for challenging her reinforces all the things we have talked about in SBTL. I am thankful for being given the opportunity to grow as a teacher and will continue to challenge my students." We do hope that this sort of student-focused reflective practice resulting in continuous change and growth will be the legacy of our project for many teachers and the students they serve.

REFERENCES

Ackerman, Cheryl M., and Ximena Uribe-Zarain. *Investigating Teachers' Instructional Leadership in the School-based Teachers Leaders Program*. Technical Report No. T2008.11.01. Newark: University of Delaware Education Research and Development Center, 2008.

Ball, Deborah L., and David K. Cohen. "Developing Practice, Developing Practitioners: Toward a Practice-based Theory of Professional Education." In *Teaching as the Learning Profession: Handbook of Policy and Practice*, edited by Linda Darling-Hammond and Gary Sykes, pp. 3–32. San Francisco: Jossey-Bass, 1999.

Erickson, Dianne. "A Problem-Based Approach to Mathematics Instruction." *Mathematics Teacher* 92 (September 1999): 516–21.

Lave, Jean. *Cognition in Practice: Mind, Mathematics and Culture in Everyday Life*. Cambridge, UK: Cambridge University Press, 1988.

McDonald, Joseph P., Nancy Mohr, Alan Dichter, and Elizabeth C. McDonald. *The Power of Protocols: An Educator's Guide to Better Practice*. New York: Teachers College Press, 2007.

Mills, Geoffrey. *Action Research: A Guide for the Teacher Researcher*. Upper Saddle River, N.J.: Prentice Hall, 2000.

Putnam, Ralph T., and Hilda Borko. "What Do New Views of Knowledge and Thinking Have to Say about Research on Teacher Learning?" *Educational Researcher* 29 (January/February 2000): 4–15.

Spillane, James, and John Diamond, eds. *Distributed Leadership in Practice*. New York: Teachers College Press, 2007.

Stazesky, Pamela B. *Measuring the Impact of the Quality of Mathematics Instruction on Student Achievement During the Middle School Years: Pilot Study*. Newark: Delaware Education Research and Development Center, University of Delaware, 2004.

Stein, Mary Kay, Margaret Schwan Smith, Marjorie A. Henningsen, and Edward A. Silver. *Implementing Standards-Based Mathematics Instruction: A Casebook for Professional Development*. 2nd ed. New York: Teachers College Press, 2009.

Thompson, Charles L., and John S. Zeuli. "The Frame and the Tapestry: Standards-Based Reform and Professional Development." In *Teaching as the Learning Profession: Handbook of Policy and Practice*, edited by Linda Darling-Hammond and Gary Sykes, pp. 341–75. San Francisco: Jossey-Bass, 1999.

Uribe-Zarain, Ximena. *School-Based Teacher Leaders: Longitudinal Analysis*. Technical Report No. T2009.17. Newark: University of Delaware Education Research & Development Center, 2009.

Part III

Teachers and Teacher Learning

What practices engage teachers in ways that support their continued learning?

Researchers in professional development have established that one-shot workshops, with a presenter telling teachers what they need to do, are not effective for improving teaching and (ultimately) students' learning. The chapters in this section underscore the importance of professional developers (e.g., teacher leaders, university faculty, school administrators) *collaborating with* teachers in determining their needs and in providing relevant and ongoing experiences that will support their work.

The following chapters describe specific forms of support and resources, telling how professional developers listened and responded to teachers' needs, and how teachers engaged in and became accountable for their learning. Ultimately, in order to positively impact their own development and their students' learning, teachers need to be empowered and positioned to have a voice and role in planning and executing professional development experiences.

As you read this section, consider the following questions:

- How can professional developers and teachers partner to design professional learning experiences?
- Once a program is underway, how can all partners work to identify needs and then adapt the program based on these needs?
- How can partners work to establish and sustain a culture of respect, professional collaboration, and accountability to improving teaching and learning?

CHAPTER 8

The Importance of Collaboration to New Teacher Development

Two Central Features of an Induction Fellowship

Rachael Eriksen Brown
Jeanne M. Vissa
Jennifer L. Mossgrove

M UCH RESEARCH in the last twenty years substantiates what many have believed: We face a national challenge of developing enough highly qualified mathematics teachers for our schools. This challenge is exacerbated by evidence that the best and brightest of new teachers with rich content knowledge leave the profession fastest of all (Henke et al. 2000). One approach for addressing this problem is to purposefully design learning opportunities for beginning teachers that support their growth as engaged professional practitioners and prospective teacher leaders. Collaboration plays a key role in this work, as research indicates that beginning teachers benefit from multiple networks of support (Wilkins and Clift 2007) and that learning is best supported within a community of practice (Hammerness et al. 2005; Ball and Cohen 1999).

The Knowles Science Teaching Foundation (KSTF) Teaching Fellowship program supports beginning high school science and mathematics teachers from the time they begin working on a teaching credential through the early years of their careers. The program was designed to retain high-quality teachers in the classroom, prepare teacher leaders, and create a learning community of people with diverse preparation who are teaching in a variety of contexts. Structures that support teachers and show the benefits of collaboration are central to the KSTF models, and are important elements of professional development for teachers in general.

Hiebert and colleagues (2007) argue that helping teachers develop the skills to critically and systematically analyze their teaching and its effects on student learning takes time, and it should therefore occur beyond the limited duration of teacher education programs. In a recent survey, 79 percent of our fellows stated that their KSTF experience strongly complements and strengthens their teacher education program. As KSTF program officers for mathematics education, we consistently reflect on and evaluate our work with fellows through structured debriefs at meetings, anonymous surveys after meetings, and a yearly survey of all fellows conducted by an outside research group. A main finding of our work—and one that relates to teacher professional development more generally—is that ongoing collaboration among our fellows is enhanced by their immersion into sustained themes over the course of the five-year fellowship. These themes are developed in whole-group activities such as meetings and discussion boards, small-group activities including lesson study, and individual but collegially supported inquiry through an annual portfolio.

After a brief overview of the KSTF fellowship, we will describe two central features—cohort meetings and the online discussion board—to explain our views about how collaboration helps to achieve our various goals for early career mathematics teacher development.

Design of the Fellowship

The KSTF teaching fellowship program supports beginning high school teachers in three disciplines—mathematics, physical science, and biology. The mathematics teaching fellowship program began in 2005, four years after the physical science teaching fellowships were first offered, and each year a new cohort of fellows is selected. The cohort of eight to fourteen KSTF mathematics teaching fellows is made up of individuals who have earned or are in the process of earning a degree with a concentration in mathematics or engineering from a recognized institution of higher education and will complete a teacher education program that leads to licensure. A third-year cohort may have fellows who are in their first, second, or third years of teaching in U.S. public and independent schools, depending on when they completed a teacher certification program. As of October 2011, the program had sixty-nine currently active mathematics fellows and former fellows from seven cohorts spread across twenty-two states.

The fellowship is designed to support the needs of beginning teachers; its features include cohort meetings and a discussion board (both described in detail later), summer and academic year professional development, membership in a professional organization, and tuition support. These fellowship benefits were selected to counteract the reasons beginning teachers most often give for leaving the profession (see Ingersoll 2003). The rest of this chapter will discuss how two central features of the KSTF teaching fellowship contribute to a professional community that supports teacher development, retention, and leadership. While each feature is presented independently here, the features are integrated within the KSTF fellowship program.

Cohort Meetings

The fellows are required to participate in three in-person meetings each year—fall, spring, and summer. The meetings are the foundation of the professional collaboration we support within and among cohorts. Some meetings are designed so that fellows can collaborate not only within their own cohort, but also across the fellowship with other mathematics cohorts, and with science cohorts as well.

Each year of the five-year fellowship focuses on a different area of knowledge for beginning teachers: (a) reconsidering mathematical content for teaching, (b) reconsidering pedagogy, (c) using assessment to monitor learning and improve instruction, (d) reaching all learners in the classroom, and (e) being a leader in mathematics education. The themes are purposefully sequenced based on our experience with induction support for new teachers. Typically, a mathematics teacher educator who has specialized in a particular theme is invited to lead a portion of the fall or spring meeting related to it. In the next few sections, we will describe the fall meeting during year two of the fellowship, as an example of how teacher collaboration is supported and developed in meetings. We discuss three facets of the meeting format critical to fellows learning from, and with, each other: yearly theme-focused activities, KSTF lesson study, and portfolio development.

Yearly Theme-Focused Activities

By the second year of the fellowship, many fellows have begun teaching in their own classrooms and a few may still be student teaching. In this year, we focus on mathematical tasks that promote critical thinking, as opposed to a general topic with little mathematical content, so that our professional development follows recent recommendations on the importance of deepening content knowledge and pedagogical content knowledge (e.g., Hill, Schilling, and Ball [2004]; Sowder [2007]). During the fall meeting, the second-year cohort typically examines qualities of mathematical tasks (see Stein et al. [2000]) and the role of the teacher in facilitating those tasks. Through activities such as solving and analyzing tasks, reading cases of practice, exploring literature on how students learn, and looking at student work, fellows have a number of opportunities to work collaboratively as they consider how teachers can plan and enact lessons that support student learning. Group participation structures during these activities include fishbowl conversations, gallery walks, and jigsaws. For example, at our most recent fall meeting for the second-year cohort, fellows worked in small groups to create high-level tasks for a given content goal. A jigsaw structure was used to share each group's task and goal while generating feedback on whether the task met that goal and was in fact a high-level task. In addition to group participation structures, we deliberately use protocols for developing shared meaning from readings assigned for a meeting, to reflect on a controversial issue, or to spread out participation and make sure key points are developed throughout a meeting.

KSTF Lesson Study

Another meeting experience that gives beginning teachers a chance to develop the skills needed for teaching while working within a community of practice is KSTF's version of lesson study. The goal of KSTF lesson study is not to produce a "polished stone" lesson, but for groups to collectively study the evolution of their understanding on the teaching of a concept. As Gorman, Mark, and Nikula (2010) advocate, we provide time and structures for fellows to engage in intellectual conversations around the teaching and learning of a specific mathematical concept. Several textbooks are made available so small groups can develop a concept map of connections that confirm, deny, or add to what they had anticipated about a topic they want to understand more deeply. We guide fellows to consider topics that span multiple courses in the secondary math curriculum, such as proportional thinking, transformations, and the inductive thinking associated with pattern interpretation. Fellows look for places where understanding might be fragile, and they develop lesson plans that function both as opportunities for probing their students' understanding further and for intervention. In this way, the Year 2 emphasis on rich task selection and on intentional questioning engages small groups of fellows in common inquiry. Fellows are asked to gather evidence of student understanding related to their topic and to use a protocol—such as the Peeling the Onion or Tuning protocols available at the National School Reform Faculty site (nsrfharmony.org)—to examine student work together for robust conceptions, partial conceptions, and misconceptions.

By grounding their lesson study work in artifacts of practice (Smith 2001), fellows are able to consider how their stated theory evolves into an enacted one. From the start of the fellowship, the engagement at our meetings models conversations and activities that fellows might initiate in their own school settings. Tools like concept maps and protocols are modified and used continuously throughout the fellowship based on the year's focus, and fellows are encouraged to critically reflect on the protocol itself. One fellow reflected, "The protocol provided a very structured approach to analyzing student work in order to make the best use of time. I learned that such a structure was needed for something as seemingly simple as looking at student work." We have found using these structures and protocols to be important components for building and supporting collaboration and community. Moreover, focusing with a variety of tools on a particular issue of mathematics content and instruction has increased our ability to sustain conversation and collaboration.

Post-meeting surveys have indicated that the fellows see the need for, and benefit of, exploring content collaboratively around a focused mathematical topic or concept. For example, one fellow wrote:

> Our knowledge of the content and standards must extend forward and backwards to see what misconceptions students may be coming in with, what skills and knowledge they are lacking, and how to best prepare them for future years of math.

Through these experiences fellows learn to base instructional decisions on evidence, to make their teaching practices public, and to take professional risks with colleagues. As

Feiman-Nemser (2001) proposed, "These skills and dispositions … are critical in the ongoing improvement of teaching" (p. 1030). Reflecting on his collaborative experience within his lesson study group, a second-year fellow stated:

> I realized the act of investing time to deliberately think about the improvement of a single lesson had an effect on every lesson that I taught. Instead of learning about lesson effectiveness through trial and error, as first-year teachers naturally do, I was embarking on a course of study to deliberately research and improve my own practices.

We expect fellows will continue to use these developing skills and dispositions regularly in their own practice and in their interactions with colleagues.

Creating Portfolios

The third recurring activity at KSTF meetings—an annual portfolio submission—addresses the common experience that beginning teachers have of feeling overwhelmed by all the elements of their practice that they could improve (Chitpin and Simon 2009). Fellows choose a dilemma of practice for the year, investigate this dilemma, and create a compelling narrative of professional growth that shows an awareness of the complexities of teaching. Although this is a personal process, fellows also use time at KSTF meetings to collaborate and work on portfolio development. KSTF portfolios depart from the typical interpretation of a portfolio as a showcase for a student's or teacher's best work. Professional teachers need to learn *how* to study their practice, as well as develop an understanding of *why* this study is important to ongoing teacher development. Our definition of *portfolio* reinforces the need for professional teachers to engage in continuous study of their practice with others. This view is aligned with those of others, such as Slavit, Nelson, and Kennedy (2009), who wrote, "When an inquiry group possesses an inquiry stance, they are more able to examine the influence of various contextual elements on their pedagogical practices and learning goals for students. . . . Through questioning, reflecting, and dialogically interacting with others, an inquiry stance positions teachers to move outside their own paradigm and explore issues of practice from different norms and perspectives" (p. 170).

Although the final portfolio is an individual product, it nonetheless benefits from collegial inquiry. Through peer collaboration, the fellows refine their initial foci as well as develop plans for studying those issues of practice. Additionally, fellows share what they are learning about their goal and the questions their inquiry is raising. Fellows spend an afternoon at both the fall and spring cohort meetings working together on their portfolios. This time for collaboration is, at times, in small teams using a protocol; at other times it occurs with the whole group participating in gallery walk feedback. At our most recent fall meeting for the second-year cohort, fellows created growth plans for their individual portfolio foci, and then followed a protocol within a small group to provide feedback on other steps that could be taken to learn about their identified dilemma of practice. We have found that the public forum of a gallery walk allows the fellows to appreciate commonalities and differences in their approaches. Responses from meeting surveys indicate that fellows value these collaborative opportunities. A fourth-year fellow commented that through the small-group discussion she was able to refine and refocus her work:

> I realized that I was concentrating [on] a totally different goal at times than I had first stated and received feedback on how I might utilize those steps further to reach my original desired goal. It is easy to get sidetracked and unfocused when working on the goal, and having feedback and accountability during the meetings and on the board helps.

The fellows' portfolio work has implications beyond the specific portfolio for areas such as developing learning communities at their school and becoming teacher leaders. For example, raising and examining an issue in teaching and studying its impact is not meant to be isolated as a single practitioner event. In the fifth year, within the theme of teacher leadership, we remind fellows that prioritizing the dimensions of a teaching issue through their initial portfolios was not a straightforward process; it was through collaboration with others that they were able to clarify and solidify their own thinking and understanding. As fellows begin to take on leadership roles in their own schools, they need to remember that many different constituencies must be involved in the deep study of a problem in order to effectively find and implement a solution for it.

Portfolio development extends into, and beyond, the fall meeting. To further support the process of portfolio development, we establish timelines for constructive dialogue on an online discussion board; this gives the process ongoing momentum as well as more opportunities for teachers to learn from one another.

Integrating the Meeting Activities

Through the meetings, fellows work within the community to continue their own learning, question beliefs, provide and receive critical feedback, and contribute to the larger field. As noted earlier, although the three facets of our meetings are presented separately, they are designed to complement each other. We also rely heavily on the online discussion board discussed below for ongoing work and discussion of annual themes, KSTF's version of lesson study, and portfolio creation. Table 8.1 shows how the plan for the Year 2 fall meeting integrates the major meeting activities with the discussion board conversations and provides opportunities for collaboration. All three facets of our meeting—yearly theme-focused activities, lesson study, and portfolios—are designed to professionalize teaching, an effort that we believe helps in retaining our fellows in the profession despite the likelihood of their leaving as reported elsewhere (Henke et al. 2000). We view the meetings as essential elements of the fellowship that anchor the annual theme, while other structures, like the discussion board, support the annual theme throughout the year.

Online Discussion Board

Collaboration on the asynchronous discussion board, where fellows are online at different times, is one of many ways fellows network within and across cohorts. The discussion board provides fellows with an online outlet for conversations about their own and others'

Table 8.1

Example of the integration of two central features in the fall of Year 2

	Annual Focus: Reconsidering Pedagogy	Portfolio	KSTF Form of Lesson Study
Prior to Fall Meeting	• **Read:** selected readings on learning theories and specific readings to prepare for fall meeting • **Discussion board post:** targeted discussion on major themes from reading	• **Discussion board post:** post goal, rationale, preliminary learning plan, and questions to small groups • Before fall meeting **discussion board post:** provide feedback to group members and respond to feedback	• Each team member collects pre-assessment evidence of student understanding • **On discussion board,** team members compare what evidence they are bringing and which textbook(s) to review at fall meeting
Fall Meeting	**Focus:** Cognitive demands of tasks	• Work in groups (different from discussion board groups): clearly articulate goal/brainstorm learning plans, provide and receive feedback	**Teams:** • Review evidence with supporting protocol • Review text according to textbook protocol • Refine enduring understandings • Plan learning opportunity and evidence to collect over this school year
Between Fall and Spring Meetings	• **Post on discussion board:** follow up to fall meeting; connect themes to research on how people learn • **Read:** research on learning theories specific to content • **Discussion board post:** targeted discussion on major themes from reading	• Submit goal, standard, rationale and learning plan to receive feedback from Program Officer • **Discussion board post:** post to small groups regarding portfolio progress, to receive and provide feedback to group members	• During the school year, team members implement learning opportunity and collect evidence • **On discussion board,** discuss preliminary impressions/findings from evidence • Each team member continues investigation of the topic

thoughts, concerns, and successes through threaded conversations, links, attached PDF and HTML documents, videos, and other artifacts related to professional growth. This board is part of a larger portal that KSTF maintains for all kinds of communication related to the fellowship. This portal currently allows fellows to file grant requests for materials or professional development/conference attendance, interact with their lesson study partners, journal, and access a library of relevant articles as well as agendas for meetings. In the following sections, we describe the logistics of the discussion board, give examples of online collaboration, and discuss our role on the discussion board. We highlight the strengths and challenges we have observed related to using the discussion board as a forum for pushing the fellows' thinking forward and building community.

Discussion Board Logistics

The discussion board is presented to fellows as a place to continue to collaborate and build community as a cohort between meetings and as a larger KSTF group. Posts on it are primarily initiated by the fellows, who are each responsible for posting at least twice per month. One standard way the board is used is to prepare for and reflect on meeting activities (see table 8.1), but it is the fellows who ultimately choose how the space is used. Some conversations focus on teaching a specific mathematics content or course component. Other posts inquire into how to use new technologies, such as interactive boards, and how to support colleagues to use them. Still other posts document the rewards and challenges of using a new practice, such as performance grading. Fellows sometimes use posts to ask for support, encouragement, and ideas for facing challenging aspects of teaching.

Our discussion board currently has strands for individual cohorts as well as specialization groups (i.e., regional groups, mathematics, announcements, and professional development opportunities). Each fellow has access to the entire board, yet they initially post primarily on their own cohort board. Fellows "favor" certain boards, meaning that they receive a notification in their personal emails of posts that have been made to those boards, along with a link that directs them to the post.

Online Collaborative Exchanges

We do encourage fellows to post across cohorts, but it is often the program officer who realizes that there may be expertise among members of another cohort for responding to a question. For example, a third-year fellow initiated a thread about "teaching taking over her life." After a program officer made the fifth-year cohort aware of the thread as a way to encourage cross-cohort collaboration and provide a leadership opportunity for more experienced fellows, one fifth-year fellow posted about the importance of maintaining a personal life:

Third-year fellow: So, fellows who have survived the "dreaded" first year, what are some tricks that you've learned to do so that you can stay sane, healthy, etc. What have you learned to give up in order to give yourself time after school? I can't wait to hear all of the sage advice ya'll have to share!

Fifth-year fellow: I remember very specifically the feeling I had driving home (after attending a midweek concert): my lesson plans for the next day may not have been the best I had ever made, but I felt re-energized.

This type of exchange encourages fellows to make dilemmas of their practice public to the community and to value the varied experiences of the group. Fellows learn to seek advice and input from other teachers instead of isolating themselves. The geographic and experiential diversity within our cohorts brings a range of perspectives to issues that program officers in a home office cannot provide alone, or, as Luebeck and Bice (2005) noted, that fellow educators in a local context cannot provide.

As prospective mathematics education leaders, fellows sometimes use the discussion board to explore the stances they might take on current or emerging issues in education. For example, one cohort recently had a conversation titled "Meeting the kids where they are at?" in which they struggled together to figure out what constitutes "readiness" for mathematics courses and tasks.

Fellow 1:   I teach at an early college high school where students immediately take algebra 1 and algebra 2 in their freshman year. I believe they have less of an attention span than they could have in two years, and they also do not take the course as seriously as the older students I have seen. It is most noticeable to me with my freshmen in algebra 2, that have a difficult time drawing general conclusions about functions and their properties. I have not done any research on this, but I was wondering: how big of an impact should maturity and age have on course sequence as opposed to completion of the prerequisite course?

Fellow 2:   I honestly think the way we force kids into these specific math classes based on age/grade groupings does more harm than good. I think in an ideal world, students would move on to the next course in the sequence when they have the requisite mathematical skills/maturity to handle the new level of complexity and abstraction, and that there would be no stigma attached to students moving at a pace that is appropriate for them.

Fellow 3:   Jo Boaler just gave an interesting talk at a conference I attended last weekend in which she referenced the "psychological prisons" from which students never escape, that is, the limits that bind students when they are told what they are capable of accomplishing, doing, demonstrating, etc. (She was talking about tracking, namely.) I don't think that this is what any of us try to do in our classrooms, but I do seriously wonder about the implicit messages that our math tasks and activities send to students and the impact they have on students and their beliefs in themselves as learners.

In this exchange, the fellows were not looking for a neat and tidy solution. Rather, they used the discussion board to compare their experiences and ideas and consider the implications for equitable teaching.

The Role of the Program Officer

KSTF has a process for monitoring and responding to online conversations that determines who (that is, a fellow's peers or a program officer) is best positioned to offer a constructive response. Simonsen and Banfield (2006), looking at student and instructor interactions in courses with online support, called attention to five interventions that should be used differentially for the greatest effect: *resolve, validate, redirect, expand,* and *withhold.* They noted that a disposition to "withhold" on the part of the instructor encourages participants to "lead and become active participants in the discussion" (p. 53), thus enhancing the mathematical discourse. We try to consider whether posing questions, suggesting resources, posting a mathematical problem, or bringing forward an issue we know the cohort is thinking about (through personal interactions or cohort meetings) will help strengthen and push the development of the KSTF community. When fellows have raised issues with program officers in private phone or email conversations that they consider more sensitive, our inclination is to work with them to make the issue public. We want to remind fellows of the value of multiple voices and opinions, and that sharing struggles and successes with a community can be a learning experience for others as well.

We have begun to have explicit debriefing conversations about the online community at our cohort meetings. Program officers initiate these conversations by posing questions to the cohort about their use of the board, the format of responses on it, and its purpose. This supports KSTF's established norms of collaboration as well as encourages fellows to reflect on how their posts can best push the thinking of other cohort members. After one of these conversations, for example, a first-year cohort engaged in expanding their content knowledge decided that the program officer should initiate a thread each month posing a mathematical task for the cohort to consider. In part, our reason for holding these conversations is to explore transitions in the profile of cohort members and the things that could change in their purposes for posting over the course of five years. "Survival" interests teachers early in their careers, but as teachers mature they begin to question the trajectory of their students' performance. As program officers, we see this as an opportunity to nurture teachers into becoming change agents in their schools regarding mathematical concepts, skills, or reasoning abilities that need to develop beyond their own classrooms.

Future Directions for the Board

The online discussion board is an essential component of KSTF's fellowship program because it extends professional development between meetings and also provides opportunities for the fellows to grow and develop as a community as they interact online. Because our fellows are located all over the country, this is essential to our success. The board also provides a safe place to publically deal with the struggles of being a beginning teacher, and a place where more experienced teachers (other fellows and program officers) can validate those struggles. Additionally, the discussion board serves as a place for the fellows to engage in and even practice conversations regarding specific situations in their schools (e.g., textbook adoption, course design, leading a small workshop for the department, or engaging parents). These conversations prompt the fellows to consider particular issues in the context of their own schools and how they might initiate similar conversations there about such issues.

Up to now, the discussion board has been the exclusive way for fellows to communicate. We acknowledge, however, that new technologies not only enable communication but also shape it, making it incumbent on us to rethink our design, practices, and purposes. For example, using older technology, fellows conversed on the discussion board about what they learned through engaging in KSTF lesson study, and appended—thereby backgrounding—the actual lesson study plan. Now we maintain wiki spaces for each group, foregrounding how activities change from iteration to iteration. Questions we are considering as we continue to modify our online support include:

- What influences who participates in conversations, and does that matter?
- Discussion board posts are typically analytical in tone; does that mean that we are missing some of the spontaneity of immediate conversations à la Twitter, Facebook, and texting?
- In a community designed to be a learning community, what recognition should there be for one-to-one communications as well as for more public communication?

Conclusion

KSTF induction support is designed to provide connected, collaborative learning opportunities for early career mathematics teachers over time by focusing on a yearly theme in multiple ways at meetings and through online discussion board exchanges. Whether engaging with mathematically rich tasks, improving a lesson, developing practice-based inquiry in a portfolio, or weighing school policies, the experiences highlighted in our induction program all focus novice mathematics teachers on student learning. As Feiman-Nemser (2001) argued,

> Unless we take new teachers seriously as learners and frame induction around a vision of good teaching and compelling standards for student learning, we will end up with induction programs that reduce stress and address immediate problems without promoting teacher development and improving the quality of teaching and learning. (p. 1031)

By the end of the five-year fellowship, we expect that each fellow will have engaged in a variety of activities and experiences that have laid a solid foundation for becoming what we term "a professional high school mathematics teacher." We believe professional development for novice teachers should include acquiring the skills and dispositions to hold courageous conversations that can maximize the opportunity for high school students' mathematics learning. Our fellows not only remain committed to teaching but also begin to take on leadership roles in their local educational communities. For example, two of our fifth-year fellows (with three and four years of teaching experience) have recently been asked to develop in-service courses on differentiation for their state. Fellows in various cohorts lead professional learning communities, while others lead professional development workshops for their departments. We believe other induction support systems—those organized by districts and other entities—need to act with the aim of achieving greater retention through developing a norm of professional collaboration that focuses on student learning.

REFERENCES

Ball, Deborah Loewenberg, and David K. Cohen. "Developing Practice, Developing Practitioners: Toward a Practice-based Theory of Professional Education." In *Teaching as the Learning Profession: Handbook of Policy and Practice*, edited by Linda Darling-Hammond and Gary Sykes, pp. 3–32. San Francisco: Jossey Bass, 1999.

Chitpin, Stephanie, and Marielle Simon. "'Even If No-One Looked at It, It Was Important for My Own Development': Pre-Service Teacher Perceptions of Professional Portfolios." *Australian Journal of Education* 53 (November 2009): 277–93.

Feiman-Nemser, Sharon. "From Preparation to Practice: Designing a Continuum to Strengthen and Sustain Teaching." *Teachers College Record* 103 (December 2001): 1013–55.

Gorman, Jane, June Mark, and Johannah Nikula. *A Mathematics Leader's Guide to Lesson Study in Practice*. Portsmouth, N.H.: Heinemann, 2010.

Hammerness, Karen, Linda Darling-Hammond, and John Bransford (with David Berliner, Marilyn Cochran-Smith, Morva McDonald, and Kenneth Zeichner). "How Teachers Learn and Develop." In *Preparing Teachers for a Changing World: What Teachers Should Learn and Be Able to Do*, edited by Linda Darling-Hammond and John Bransford, pp. 358–89. San Francisco: Jossey-Bass, 2005.

Henke, Robin, Xianglei Chen, Sonya Geis, and Paula Knepper. *Progress through the Teacher Pipeline: 1992-1993 College Graduates and Elementary/Secondary School Teaching as of 1997* (NCES 2000-152). Washington, D.C.: U.S. Department of Education, 2000.

Hiebert, James, Anne K. Morris, Dawn Berk, and Amanda Jansen. "Preparing Teachers to Learn from Teaching." *Journal of Teacher Education* 58 (January/February 2007): 47–61.

Hill, Heather C., Stephen G. Schilling, and Deborah Loewenberg Ball. "Developing Measures of Teachers' Mathematics Knowledge for Teaching." *The Elementary School Journal* 105 (September 2004): 11–30.

Ingersoll, Richard M. *Is There Really a Teacher Shortage?* Seattle: University of Washington, Center for the Study of Teaching and Policy, 2003.

Slavit, David, Tamara Holmlund Nelson, and Anne Kennedy. "Supporting Collaborative Teacher Inquiry." In *Perspectives on Supported Collaborative Teacher Inquiry*, edited by David Slavit, Tamara Holmlund Nelson, and Anne Kennedy, p. 170. New York and London: Routledge, 2009.

Luebeck, Jennifer L., and Lawrence R. Bice. "Online Discussion as a Mechanism of Conceptual Change among Mathematics and Science Teachers." *Journal of Distance Education* 20 (2005): 21–39.

Simonsen, Linda, and Jeff Banfield. "Fostering Mathematical Discourse in Online Asynchronous Discussions: An Analysis of Instructor Interventions." *Journal of Computers in Mathematics and Science Teaching* 25 (January 2006): 41–75.

Smith, Margaret Schwan. *Practice-Based Professional Development for Teachers of Mathematics.* Reston, Va.: National Council of Teachers of Mathematics, 2001.

Sowder, Judith T. "The Mathematical Education and Development of Teachers." In *Second Handbook of Research on Mathematics Teaching and Learning,* edited by Frank K. Lester, pp. 157–223. Charlotte, N.C.: Information Age Publishing, 2007.

Stein, Mary Kay, Margaret Schwan Smith, Marjorie A. Henningsen, and Edward A. Silver. *Implementing Standards-Based Mathematics Instruction: A Casebook for Professional Development.* New York: Teachers College Press, 2000.

Wilkins, Elizabeth A., and Renee T. Clift. "Building a Network of Support for New Teachers." *Action in Teacher Education* 28 (Winter 2007): 25–35.

CHAPTER 9

Building a School-University Collaboration

A Search for Common Ground

Margaret S. Smith
Jennifer L. Cartier
Samuel L. Eskelson
Miray Tekkumru-Kisa

C ROSSING THE BOUNDARIES between universities and schools presents significant challenges for collaborators. Teachers often perceive university researchers to be unaware of the realities of the classroom, unwilling to maintain relationships over the long haul, and "users and abusers" of their time and their students (Barnett, Higginbotham, and Anderson 2006). Institutional differences and priorities can result in conflicts among participants and make it difficult to establish key features of successful collaborations such as shared goals, adequate leadership structures, mechanisms for resource allocation, and agreed-upon work products (Thorkildsen and Scott Stein 1996). In this chapter we argue that university partners must learn how to bridge the gap between the university and school contexts by engaging in collaborative work directly with classroom teachers. Moreover, we argue that university partners must be flexible and willing enough to adapt their professional development activities to meet teachers' needs for support that is closely related to their daily work (Stein, Smith, and Silver 1999).

Collaborations between schools and universities can have different purposes and structures. Handler and Ravid (2001) describe such models for school-university collaboration as the Umbrella model, where "multiple collaboration project teams" operate within "one umbrella organization acting as the facilitator" of the work (Handler and Ravid 2001, p. 4). The multiyear Lesson Planning Project (Stein et al. 2008) that we describe in this chapter is a collaboration in which discipline-specific teams conduct weekly professional development activities in College Ready, a local secondary school for grades 6–12, under the umbrella of a university-based urban education center (fig. 9.1). A core premise of the Lesson Planning Project was the belief that teachers' engagement in thoughtful, thorough lesson planning routines would lead to more rigorous instruction and improved student learning. Thus, the researchers' major goals were to help teachers enact particular lesson planning practices using tools adapted from products of prior research (Cartier and Pellathy 2009; Smith, Bill, and Hughes 2008). (The researchers also worked toward some other goals and engaged in collaborative work with administrators, but our focus here is on their work with teachers.)

The voice in this chapter is that of four of the university researchers/teacher educators involved in the Lesson Planning Project. Here we tell the story of how we adapted our project activities to take advantage of some school contextual factors (while minimizing the influence of others) in order to move closer to achieving our ultimate goal: having an impact on students through the actions of teachers. Central to our story is our search for common ground—the struggle to connect *our* agenda (a set of research-based ideas related to teaching and learning as embodied by planning practices and routines) with what mattered to teachers (concrete suggestions for improving student learning)—in order to improve the practice of teaching and learning at College Ready.

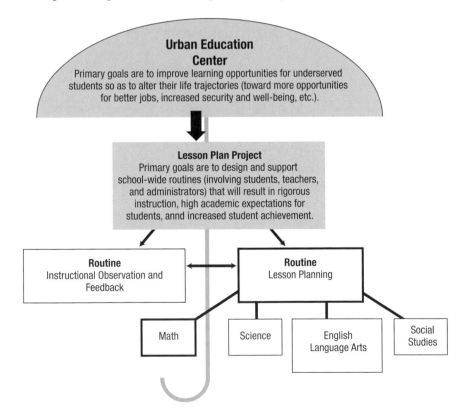

Fig. 9.1. Overview of the Lesson Planning Project operating under the umbrella of the Urban Education Center

The Project

The Lesson Planning Project (Stein et al. 2008) aims to design and study a set of routines and tools anchored in lesson planning for promoting schoolwide change and instructional improvement. Launched in the fall of 2009, the project is housed within a partnership between an urban education center at a large urban research university and the city school district, as shown in figure 9.1. The university-based team includes experts in organizational and school reform as well as instruction in mathematics, science, social studies, and English language arts. Although the project work includes these four subject areas, the focus of this chapter is on the work of the mathematics team (Stein, Russell, and Smith 2011).

To achieve the project's objectives related to enhancing lesson planning and the conversations around it, a broad array of activities were undertaken at the project site, including (1) the development and refinement of an electronic lesson planning tool for teachers that is based on the Thinking Through a Lesson Protocol (Smith, Bill, and Hughes 2008); (2) co-planning meetings of university-based instructional experts (UBs) and teachers; and (3) co-observations of classrooms by university-based partners and school principals along with debriefing and feedback sessions with teachers.

The Setting

College Ready is a 6–12 public school primarily serving African American students from low-income neighborhoods. The school is housed in a former district middle school that had been closed for several years due to declining enrollments. For the school year of 2008–2009, College Ready opened its doors to a small group of ninth-grade students, and in fall 2009 the school dramatically grew to include students ranging from sixth grade to tenth grade. In the fall of 2010 there were 528 students enrolled in the sixth through eleventh grades; by the fall of 2011 the school will be running at its full capacity of 650 students in grades 6–12.

The school day at College Ready is structured in eighty-minute blocks for core subjects (math, language arts, science, and social studies), and teachers generally are assigned three instructional blocks. In addition, each day teachers have a planning period and a professional development period (forty minutes each) that is intended to provide them with time to engage in activities that will help them to continue to grow and develop as professionals. For the core subject teachers, this includes a weekly meeting with UBs.

In the fall of 2009, there were seven mathematics teachers at College Ready. (Four special education teachers also either provided in-classroom support or pulled students out in small groups for mathematics instruction; these teachers were not required to go to the co-planning meetings and attended them only sporadically.) The seven teachers included David Ingersol, a mathematics coach who taught one class and provided support to the teachers, and Olivia Nelson, the Instructional Team Leader who taught two classes and also served as a resource. (All teachers in this chapter have been given pseudonyms.)

The teachers varied with respect to their number of years in teaching (one to fourteen) and their area of certification (elementary K–6; mathematics 7–9; mathematics 7–12). Most of the teachers applied to College Ready because they had been "displaced" from their previous position in the district as a result of school closures or else were dissatisfied with their previous assignment. Although there was some overlap in teachers' assignments at certain points during the year (e.g., two teachers taught geometry in the fall; two other teachers taught algebra 2 in the spring), most teachers did not consistently have a colleague teaching the same content.

The UBs consisted of an experienced mathematics teacher educator (TE) and a graduate student (GS) in mathematics education who had previously taught high school mathematics (the first and third authors of this chapter respectively). The TE had worked with secondary mathematics teachers in various projects and classes for nearly two decades, and she had spent many years working with teachers in the district in which College Ready 6–12 was located.

Co-Planning Meetings at College Ready

The UBs drew upon the theory of change depicted in figure 9.2 (and adapted from Desimone [2009]) to create the design of their co-planning routines with teachers. As shown in the figure, professional development (in the form of co-planning meetings) was intended to improve teachers' knowledge, which in turn would influence the ways in which they planned for and enacted instruction. Changes in classroom instruction would ultimately lead to improved student-learning outcomes.

Fig. 9.2. Framework for linking the effects of professional development on teachers and students

The UBs initially decided to focus the professional development on two dimensions of teachers' knowledge and related pedagogical skill: (1) the design and selection of instructional tasks, and (2) classroom talk. The focus on tasks and talk made sense, given that the rigor of instructional tasks and the nature of discourse have been identified as two of the critical dimensions of classrooms that promote understanding (Carpenter and Lehrer 1998; Hiebert et al. 1997). Additionally, both had been identified as district-level priorities in mathematics. Thus, the UBs' goal in the co-planning activities was to promote the use of cognitively challenging tasks that, first of all, would engage students in thinking, reasoning, and problem solving (Stein et al. 2009) and, secondly, would result in productive classroom talk that made student thinking and reasoning public so that it could be refined or extended (Stein et al. 2008). The plan was for teachers to initially analyze artifacts (e.g., tasks and vignettes based on authentic practice) drawn from "someone else's classroom" so that they could learn how to identify good tasks and talk and their impact on student learning, and also draw connections to their own practice (Smith 2001; Ball and Cohen 1999). Ultimately, the intent was to transition to discussions that focused on artifacts from the teachers' own classrooms and to engage teachers in collaborative planning.

The goals that the UBs held for the co-planning meetings did not change over the eighteen months of the project discussed in this chapter, but the design of the meetings went through four iterations in an attempt to find a format that aligned what the UBs thought teachers needed with what was important to the teachers themselves. In the next four sections we will consider the design at each of four points between fall 2009 and fall 2010. For each point, we will discuss the meetings' primary focus and rationale, and how much their design appeared to accomplish what was intended. In telling this story of our work at College Ready we draw on field notes, artifacts, and transcripts from the co-planning meetings and interviews with teachers at the beginning and end of the 2009–2010 school year.

Design 1: Accountable Talk for All (Early Fall 2009)

Early in the fall term the UBs realized that, in order to achieve the larger goal of promoting co-planning and the sharing of personal artifacts related to teaching practice, it would be necessary to build this diverse group of teachers into a community. Community building meant listening to teachers' needs and what they thought was important (Sztajn et al. 2007) as well as providing resources when possible. This included providing teachers with basic materials (such as chart paper, markers, and rulers) and instructional resources (ancillary print and electronic resources, including specifically designed applets) and helping teachers as needed (for example, assisting in scoring quarterly assessments and acting as intermediaries with the district regarding materials and policies). To learn more about the teachers' perceived needs, the TE had a conversation with Olivia Nelson. As Instructional Team Leader, Ms. Nelson saw herself as a resource and support for teachers in the building and as an expert. The TE asked Ms. Nelson what she thought would be a productive focus for co-planning meetings with teachers. Ms. Nelson said that Accountable Talk (Resnick, Michaels, and O'Connor 2010; Chapin, O'Connor, and Anderson 2009) would be an appropriate starting point, as it had been identified as a district priority. From the UBs' perspective, this aligned well with what they had planned to do and, by adopting Ms. Nelson's framing of the work, they sent the message that they were willing to listen and to accommodate what she saw as important.

During one of the first sessions, the TE selected vignettes (shown in fig. 9.3) that had been developed as a product of prior research and asked the teachers to analyze them to determine whether the talk in either classroom was accountable to rigorous thinking. The vignettes were chosen to illustrate particular features of mathematical instruction, and they were also intended to be short enough so that teachers could read and discuss them within the forty-minute session. In the case of Mr. Johnson's vignette, the task itself was not rigorous, while in the case of Ms. Arnold, the task was rigorous but students were not held accountable for discussing the mathematics. By focusing the teachers' attention on these two vignettes, the TE intended to help them realize that engaging students in talk (which both Johnson and Arnold did) was not sufficient—the talk had to bring forth student thinking about mathematical ideas (which did not occur in either case).

Mr. Johnson's Seventh-Grade Class

Students in Mr. Johnson's seventh-grade class were completing a task that required them to express various ratios, presented in a variety of formats (e.g., 4:12, 15/25, 1/5 to 1/2, and in verbal problems), in their simplest terms. Mr. J modeled the solution of a few examples and then, in order to have students experience mathematics as a collaborative activity, he encouraged students to work and talk with each other in small groups. As they worked, he circulated around the room, stopping periodically to ask questions. When most of the class had completed the assignment, Mr. J orchestrated a large-group discussion to review the solutions. In order to promote thoughtful classroom discourse, Mr. J frequently asked students questions about their answers (e. g., "How do you know?" "Does it make sense?" or "Can you justify your reasoning?") and engaged the entire class in the discussion by asking other students to add to what had been said, whether they agreed with what had been said, or to put an explanation in their own words. (excerpt from Silver and Smith 1997)

Ms. Arnold's Sixth-Grade Class

Many of the students in Ms. Arnold's class were not native English speakers and were somewhat self-conscious and nervous about making public presentations. The first unit of study in the curriculum dealt with data representation. After students were familiar with several data representation techniques (e.g., bar graphs, line graphs, pictographs), Ms. A asked them to survey classmates about their favorite in a category of their own choosing (e.g., television shows, musical groups). Students worked in groups of four to collect data from classmates and others in the school, after which they made a graphical display of their findings. Once the graphs were completed, Ms. A asked each student group to select a spokesperson who would make a presentation to the class. After each presentation, students in the class were invited to ask questions of the spokesperson, in order to clarify their understanding.

Fig. 9.3. Vignettes used in teacher professional development at College Ready

In addition to satisfying several mathematical goals, Ms. A intended to use this project as an experience that would help her students become comfortable with public presentation and discussion. In order to ensure safe participation, she gave clear ground rules for the discussion, including making it clear that no disrespect or ridicule would be tolerated. In this regard, the ensuing discussion was a great success. Students listened carefully to each presentation, and in their questions and comments they were quite respectful of each other. Students' questions tended to deal with nonmathematical aspects of the design process (e.g., "How did you decide which TV shows to include?" "How long did it take to design the graph?" "Why did you use yellow to represent *The Simpsons*?" "Why did you decide to use to use colored pencils to draw your graph?"). (excerpt from Stein et al. 2009)

Fig. 9.3—*Continued*

Although the vignettes did generate some discussion, there was limited participation and, despite the TE's efforts, teachers were not focusing on the aspects of the vignettes that mattered most—that the classroom talk was not providing insights into how students were thinking about rigorous content. Instead, with regard to Mr. Johnson, teachers focused on how he arranged students to work on the task (e.g., beginning with smaller groups and moving to larger groups) and on the ownership of the solutions (i.e., who produced the solutions, Mr. Johnson or his students). In discussing the case of Ms. Arnold, teachers made note of the issues she had to face because of the larger number of nonnative English speaking students. In neither case did teachers substantively discuss the mathematical aspects of the tasks or the manner in which students engaged in them.

As the weeks passed, teachers became less engaged in the co-planning sessions, showing up late, leaving for long periods of time, and, in one case, sitting on the periphery of the group and electing not to participate. Some teachers felt that, because the sessions did not focus specifically on what they were teaching, they were not relevant, while a few others felt they had done it all before. A review of meeting notes suggests that few teachers were making any public contributions to the discussion and only one teacher made public connections between the ideas being discussed and her own practice.

Design 2: Subject-Specific Planning in Pairs (Late Fall 2009/ Early Winter 2010)

Given the obvious lack of enthusiasm for the focus of the co-planning sessions, and the lack of accountability for doing anything more than showing up, the UBs knew that a change was needed. The new design was intended to focus directly on lesson planning (thus eliminating the challenge of asking teachers to connect the pedagogical themes within the vignettes with their own practice) and to separate middle and high school teachers into two groups for this work. (The high school teachers are the focus of the remainder of the design discussion, as their meetings were better documented.) High school teachers who taught at least one section of the same course were paired together to engage in collaborative planning, resulting in an algebra team and a geometry team. This design was intended to center the co-planning discussions on tasks that teachers were going to enact in their own classrooms. Each session included time for teacher pairs to work collaboratively on a lesson and also time for group discussion and feedback.

Within a few weeks of instituting the new plan, it became clear it was not working as intended. Neither team was working well together despite the UBs' efforts to intervene (each UB was working with one pair). The teachers were either unable or unwilling to even find a task that they could agree to work on (the algebra team) or hopelessly deadlocked on the approach to take in enacting the task (the geometry team). Finally the UBs decided that the high school team would break into two groups—the two teachers who wanted to co-plan a lesson (Group A) and the two teachers who did not (Group B). Group A, plus the mathematics coach Mr. Ingersol, continued to meet with the UBs. Group B met with Mr. Ingersol only. Each new group included one teacher from the initial geometry team and one teacher from the initial algebra team. (At times Group A was joined by the "technology guys"—two university partners providing support for the use of computers—who gave suggestions on how technology could help in the learning of specific content and developed applets for use during instruction.)

The UBs engaged Group A in planning an algebra lesson that Zena Davis would teach, as she had expressed a need for assistance in dealing with her algebra class. The team offered feedback on the lesson plan that Ms. Davis and her colleague, Ms. Edwards, created, and most of the members of the team attended at least one day of the three-day lesson when it was taught. A lesson debriefing was held during which the two teachers in Group A, Mr. Ingersol, the "tech guys," and the UBs provided comments in the form of what they "noticed" and what they were "wondering about," a structure for framing feedback that was derived from the TE's earlier research (Smith 2009; Hughes et al. 2009). *Noticings* are intended to be factual, nonevaluative statements of an observable phenomenon. *Wonderings* are questions about some aspect of teaching and learning that are framed as inquiries (e.g., I am wondering why, how, when) intended to provoke reflection on decision making. During this debriefing meeting, multiple team members *noticed* that the students in Ms. Davis' class varied greatly in their experience, ability, and mathematical knowledge and, as a result, some students had trouble figuring out how to even approach the task, while others completed the task by the end of the second day. They *wondered* what instructional strategies Ms. Davis could employ to enable such a broad range of students to access the task and participate productively. The group decided that Ms. Davis should choose another task she would be teaching her class several weeks later, and they would focus on how to structure the task to provide suitable entry points for both the advanced students and those who were struggling.

For the two teachers in Group A, focusing on the practice of one member of the team seemed to be a fruitful pathway for developing pedagogical knowledge. Unfortunately, the two high school teachers in Group B were no longer attending the co-planning meetings with the UBs, and it is unclear whether their meetings with Mr. Ingersol were occurring on a regular basis or if they were productive when they did take place. At this point, the UBs acknowledged that the new co-planning design was not working for all teachers and needed to be revised. However, before any changes could be made, the co-planning meetings were abruptly terminated in mid-January so that teachers could turn all of their attention to preparing students for the state assessment that would be given in April. Ms. Nelson, the instructional team leader, had terminated the meetings without authorization from the school's principal; when the principal became aware of the situation, professional development was reinstated and teachers were told that they if did not want to work with the UBs they should plan to find a job in a different school within the district.

Design 3: High School Teachers Focus on Tasks (Spring 2010)

Several weeks after the unexpected schedule upheaval, the co-planning sessions began again. The UBs introduced a new structure to the co-planning sessions in an attempt to involve all of the high school teachers in discourse directly related to their practice. Each week one of the five high school teachers (including the coach) would identify a task they would be using in an upcoming lesson. This task would form the basis of discussion at the co-planning meeting. During the session the other teachers recorded what they *noticed* and what they were *wondering* about with respect to the task. Teachers were also prompted to write questions they thought the focus teacher should ask her students during the lesson. (This was actually rarely completed by teachers or discussed in the co-planning meetings.) Following a brief period of writing, each teacher shared his or her noticings and wonderings aloud. Writing comments about their colleagues' tasks added to the teachers' active engagement in the co-planning sessions and resulted in a tangible record of their participation for accountability purposes.

This third co-planning design prompted markedly different participation from the teachers. For example, at the end of April, Zena Davis shared the Absolute Value Functions task, which is summarized in figure 9.4. During this meeting, four of the five teachers identified something that they *noticed* about the task and each of the teachers identified at least two things they were *wondering* about. For the most part, the noticings highlighted features of the task, such as its length and what students were being asked to do (e.g., *compare functions to the parent function, use multiple representations*). Wanting to push teachers to think critically about the cognitive demands of the task, the TE shared her noticing that the task only prompted students to *describe* what happened in the transformations, not to *make sense of* what they observed. She went on to wonder whether it would be helpful to ask students to explain why the transformation worked as it did. Teacher wonderings focused on the availability of resources (e.g., *"Will students have technology available?"* *"Will you provide different colors?"*), meeting the needs of learners (e.g., *"How will you make accommodations for students who struggle with the task or become frustrated?" "How will misconceptions be brought out?"*) and outcomes (e.g., *"What is the learning residue you are looking for?" "How are you planning on presenting and checking for student understanding?"*).

Absolute Value Functions (AVF)

a) $y = |x|$ (defined as the AVF)
b) $y = |x - 2|$
c) $y = |x - 3|$
d) $y = |x| - 3$
e) $y = |x| + 2$
f) $y = 4|x|$
g) $y = 0.5|x|$
h) $y = -|x|$

For each function: Complete a table of values, graph the function, and identify the x and y intercepts, the coordinates of the vertex, and the slopes of the two sides of the functions. For functions b–h, sketch the parent function (a) on the same grid using a different color and compare the intercepts, the vertex, the shape, and the slope of the parent function, and the transformation. Compare function pairs (b and c; d and e; and f and g) and explain how the transformation (e.g., adding inside the absolute value function) changes the parent function.

Given a function that combines two transformations, $y = 0.25|x - 4|$ and $y = -2|x + 4| + 1$, predict how the graph of the transformed function will compare to the parent function, and then check your predictions by completing the table and graphing the function.

Fig. 9.4. The task discussed by the high school team in April

In general, the teachers' wonderings identified key issues that would be important for the teacher to consider before teaching the lesson. For the most part, however, they did not address the level and kind of student thinking that would be elicited from the task. The one exception to this was a noticing and wondering offered by the TE. She noticed that the task began by defining what the AVF was and wondered if it would make sense to first give students ordered pairs so that they could graph the parent function and then subsequently make observations about what the function did. Her intent was to subtly raise the issue of how the cognitive demands of a task can be increased by restructuring it to avoid telling students key ideas and to instead enable them to explore patterns themselves and draw conclusions from their exploration.

Design 3 was viewed favorably by three of the teachers (the former Group A and Mr. Ingersol) but drew some ambivalence from the former members of Group B. For example, at the end of the Year 1 of the Group B teachers, Olivia Nelson commented:

> Well, right now we're talking about people's activities or tasks that they're doing. I think in some sense it's good, but it's not beneficial to me as much as it is to my colleagues where they are teaching something for the very first time. As I've taught it before, I may not have taught it this particular way, but I've taught it before so I've got a lot of tricks in my hat that I can pull out whereas they don't have that option; so having another way look at it. . . . So I think it's helpful to them, but I wouldn't call it beneficial to myself, not right now.

By contrast, at the end of the year, in response to the question "Are there ways in which these meetings have supported your work this year?" Nancy Edwards (from Group A) responded:

> I would say that especially since we've done the noticing and wonderings, I've done it for two activities now, and for both of them it brought up some ideas that I actually tried and I liked the way they worked out. So having more brains to look at it and help critique the whole "Why do you try it this way?" might help more, especially since I am only a second-year teacher, but I think I'm doing pretty darn good for that. But I can get more experienced teachers to give me more ideas because they have more experience with a wider range of kids, so that has helped me significantly. I like that. It also gives me a chance to just see some of the other math teachers. . . .

Taking Stock of Effective Design Features: Reflecting on Year 1

By the end of the first year of the Lesson Planning Project the frequency of substantive participation by the math teachers in co-planning activities had increased. The increase resulted from certain features of the design: teachers had to take turns presenting a task to their colleagues, and they had to produce written feedback during each co-planning session. This design was sanctioned by the principal and did not allow teachers to elect nonparticipation. Moreover, the noticing and wondering tool helped to promote more substantive discourse about practice, drawing teachers' attention to features of instructional tasks and encouraging them to consider how to alter tasks to increase the cognitive demand for students. Still, the UBs felt that discourse would improve if teachers were discussing shared instruction rather than tasks only one member of the team intended to enact. In addition, the UBs felt constrained by the limited time that was available (it took five weeks to complete a full cycle of sharing because the team met only once a week for forty minutes).

Concurrent with our reflection about Year 1, the College Ready's parent district adopted a new system of teacher evaluation based on the Danielson framework (2007). We will refer to this framework as the New Evaluation System (NES). NES focused on collecting evidence of teacher proficiency in four domains—planning and preparation, classroom environment, teaching and learning, and professional responsibility. Each domain included a set of skills, a subset of which had been targeted for focus in the first year of NES adoption by the district. These "power components" would form the basis of teacher evaluation in the coming year, and each teacher would have an opportunity to gather evidence related to his or her proficiency in these areas. For example, within the domain of teaching and learning, *using questioning and discussion techniques* and *engaging students in learning* were identified as power components.

From the point of view of the UBs, the NES was well aligned with what they had been trying to accomplish the previous year. The overarching focus of the NES was on planning for rigorous instruction that engaged students in thinking and reasoning, and that was exactly what the UBs had been trying to address through their work on tasks and talk. The UBs therefore decided to leverage the existing accountability system represented by NES to motivate teachers' participation in co-planning meetings in the coming year.

Design 4: Subject-Specific Planning in Small Groups Revisited (Fall 2010)

Because the number of teachers had increased from seven to ten with the addition of eleventh-grade teachers and the inclusion of a few special education teachers, the decision was made to group the teachers by content or grade level. This would make it possible to better meet the needs of individual teachers and to bring the discussion closer to teachers' practice. The result was four teams of teachers—two teams with two members and two teams with three members—who taught the same content as at least one other person in the group. (The only exception to this was Zena Davis, the only teacher currently teaching geometry, who was grouped with two other teachers, one of whom had taught geometry the previous year.)

The groups began by identifying the power component that they wanted to work on during the first quarter. Each team then worked on a two-week cycle: During Week 1 members of the team met without UBs and discussed a specific lesson they planned to teach that week with particular attention to the power component they were targeting. For example, if the group was working on *using questioning and discussion techniques*, they would focus explicitly on the questions they would ask that would require students to think, problem-solve, and defend conjectures and opinions. They would use the electronic lesson-planning tool to record the changes they had made to the plan. The UBs asked teachers to collect evidence from the implementation of the lesson (e.g., audio or video recording of the lesson, copies of student work) for discussion during the following week. During Week 2, members of the team met with UBs to discuss the specific lesson that was taught the previous week. The discussion was driven by what teachers planned to do (as reflected in the electronic planning tool) and what they actually did do (as reflected in the evidence they collected). The members of the team engaged in a round of noticing and wondering about the work that was presented. In addition, the UBs asked teachers to produce at least one written reflection monthly, using the electronic lesson-planning tool.

As of the writing of this chapter, it is too soon to say whether this design will have the desired impact—more collaborative work, better planning, improved teaching, and increased levels of student achievement—we have seen some hopeful signs. Teams have co-planned lessons; they are collecting evidence (photos and videos of students working in their classrooms, and samples of student work) to show that students are engaged in rigorous mathematical activity; and two teachers have posted written reflections using the electronic lesson-planning tool that provide some insight into their teaching practice. For example, one teacher commented:

> I had students working on the Analyzing Bar Graphs task where they were looking at intervals of college tuitions for 32 different universities in the US. I anticipated four different solution paths. All of which came out in my first period. I expected some students to take $32/2 = 16$ to find the median interval, to make the graph into a frequency table and find the median by hand, to write out the tuitions by hand and solve for the median, and also to use the calculator. All students were very engaged in the task, and even students [who] usually do not participate were working well with their peers. I had students able to explain their reasoning. One video has a student explaining how he found the median using the "dividing in two" method. I have another video of a group of two boys and a girl working, and when they explained their reasoning, she asked a question. When I prompted them to answer her question and to include her, they proceeded to try to explain to her what they meant by "find the median of the median."

In this reflection we see that the teacher actually anticipated before the lesson how students would solve the task, collected evidence during the lesson to support her claims regarding students' ability to explain their thinking, and held students accountable for working with their peers. Finally, after twelve months, we had a window into what teachers were doing that would provide a solid grounding for discussions about teaching and learning that can lead to improved practices.

Conclusion

Stein and her colleagues (Stein, Smith, and Silver 1999, p. 237) argue that transforming teaching and learning in our nation's classrooms will require that teacher educators learn to provide assistance to teachers "in new settings in new ways." This means moving beyond simply offering university courses and workshops to willing participants to taking on the challenges of working with groups of teachers in school settings. The UBs described herein took on this challenge as they sought to find a design that would allow them to challenge teachers' "prevailing practices and beliefs" (Stein et al. 1999, p. 237) about the nature of tasks, talk, and lesson planning so as to improve the learning outcomes for College Ready students.

In reflecting on the design of the work with teachers over the year, several elements appear to be critical. The first element is connecting co-planning meetings to the daily work of teachers in a concrete and explicit way. While the UBs had from the beginning focused on the work of teaching, it became clear that teachers would only see real value in the work if it focused on what *they* were doing (and not on what some other unknown teacher or even one of their colleagues was). Second, the tools provided structures that supported the work. For example, the noticing and wondering framework helped to structure conversations about practice in ways that gave all teachers an entry into the discussion and held them accountable for engaging in it. The electronic lesson-planning tool provided a way to work collaboratively with partners on creating lesson plans and provided a vehicle for reflecting on practice. While the electronic lesson-planning tool had been available since the beginning of the project, it was only in the final design that some teachers began to see it as a useful resource that supported their efforts. Third, connecting the co-planning work with the NES accountability system gave the work legitimacy. That is, teachers came to see the work of co-planning as aligned to district priorities, and they realized that engaging with UBs in this effort would help them to develop as professionals in ways that mattered to them (and to the administrators who evaluated them). Finally, while the UBs may be seen primarily as teacher educators in this context, the researcher's stance of inquiry they adopted while working with teachers was a key factor in the design process. That is, because the UBs were continually seeking to understand what was or was not working, the design of the experiences evolved and flexed as was needed for the teachers' learning.

While it is too soon to say whether or not the project will ultimately have the desired impact on teachers, instruction, and learning, the takeaway lesson here is that in order to get research-related ideas and routines into classrooms, researchers/teacher educators have to be opportunistic and willing to evolve. We have to find a balance between causing discomfort and easing tension; we have to position our work in ways that don't alarm or alienate teachers and whenever possible leverage existing routines to get our work done.

......

Work on this paper was supported by the "Collaborative, Technology-Enhanced Lesson Planning as an Organizational Routine for Continuous, School-Wide Instructional Improvement" Project, which is funded by a research grant from the Institute for Education Sciences (Award Number: R305A090252). All opinions and conclusions expressed in this paper are those of the authors and do not necessarily reflect the views of any funding agency.

REFERENCES

Ball, Deborah L., and David K. Cohen. "Developing Practice, Developing Practitioners: Toward a Practice-Based Theory of Professional Education." In *Teaching as the Learning Profession*, edited by Linda Darling-Hammond and Gary Sykes, pp. 3–32. San Francisco: Jossey-Bass, 1999.

Barnett, Michael, Thomas Higginbotham, and Janice Anderson. "Didn't I Tell You That?: Challenges and Tensions in Developing and Sustaining School-University Partnerships." *International Conference on Learning Sciences* (2006): 23–29.

Carpenter, Thomas P., and Richard Lehrer. "Teaching and Learning Mathematics with Understanding." In *Mathematics Classrooms That Promote Understanding*, edited by Elizabeth Fennema and Thomas A. Romberg, pp. 19–32. Mahwah, N.J.: Lawrence Erlbaum Associates, 1999.

Cartier, Jennifer L., and Stephen Pellathy. "Integration with Big Ideas in Mind." *Science & Children* (November 2009): 44–47.

Chapin, Suzanne H., Catherine O'Conner, and Nancy C. Anderson. *Classroom Discussions: Using Math Talk to Help Students Learn, Grades 1–6*. Sausalito, Calif.: Math Solutions, 2003.

Danielson, Charlotte. *Enhancing Professional Practice: A Framework for Teaching (Second Edition)*. Alexandria, Va.: Association for Supervision and Curriculum Development, 2007.

Desimone, Laura M. "Improving Impact Studies of Teachers' Professional Development: Toward Better Conceptualizations and Measures." *Educational Researcher* 38 (April 2009): 181–99.

Handler, Marianne, and Ruth Ravid. "Models of School-University Collaboration." In *The Many Faces of School-University Collaboration: Characteristics of Successful Partnerships*, edited by Ruth Ravid and Marianne Handler, pp. 3–10. Englewood, Colo.: Teacher Ideas Press, 2001.

Hiebert, James, Thomas P. Carpenter, Elizabeth Fennema, Karen C. Fuson, Diana Wearne, Hanlie Murray, Alwyn Olivier, and Piet Human. *Making Sense: Teaching and Learning Mathematics with Understanding*. Portsmouth, N.H.: Heinemann, 1997.

Hughes, Elizabeth K., Margaret S. Smith, Melissa D. Boston, and Michael Hogel. "Case Stories: Supporting Teacher Reflection and Collaboration on the Implementation of Cognitively Challenging Mathematical Tasks." In *Inquiry into Mathematics Teacher Education*, edited by Fran Arbaugh and P. Mark Taylor, pp. 71–84. Monograph Series, Volume 5. San Diego: Association of Mathematics Teacher Educators, 2009.

Resnick, Lauren B., Sarah Michaels, and M. C. O'Connor. "How (Well-Structured) Talk Builds the Mind." In *Innovations in Educational Psychology: Perspectives on Learning, Teaching, and Human Development*, edited by David D. Preiss and Robert J. Sternberg, pp. 163–94. New York: Springer, 2010.

Silver, Edward A., and Margaret S. Smith. "Implementing Reform in the Mathematics Classroom: Creating Mathematical Discourse Communities." In *Reform in Math and Science Education: Issues for Teachers*. [CD-ROM]. Columbus, Ohio: Eisenhower National Clearinghouse for Mathematics and Science Education, 1997.

Smith, Margaret S. *Practice-Based Professional Development for Teachers of Mathematics*. Reston, Va: National Council of Teachers of Mathematics, 2001.

—————. "Talking About Teaching: A Strategy for Engaging Teachers in Conversations About Their Practice." In *Empowering the Mentor of the Beginning Mathematics Teacher*, edited by Gwen Zimmermann, Patricia Guinee, Linda M. Fulmore, and Elizabeth Murray, pp. 33–35. Reston, Va.: National Council of Teachers of Mathematics, 2009.

Smith, Margaret S., Victoria Bill, and Elizabeth K. Hughes. "Thinking Through a Lesson Protocol: A Key for Successfully Implementing High-Level Tasks." *Mathematics Teaching in the Middle School* 14 (October 2008): 132–38.

Stein, Mary K., Jennifer L. Russell, Louis Gomez, and Kimberley Gomez. *Technological-Enhanced Lesson Planning as an Organizational Routine for School Improvement*. Submitted to Institute for Education Sciences Grant Competition, 2008.

Stein, Mary K., Margaret S. Smith, and Edward A. Silver. "The Development of Professional Developers: Learning to Assist Teachers in New Settings in New Ways." *Harvard Educational Review* 69 (Fall 1999): 237–69.

Stein, Mary K., Randi A. Engle, Margaret S. Smith, and Elizabeth K. Hughes. "Orchestrating Productive Mathematical Discussions: Helping Teachers Learn to Better Incorporate Student Thinking." *Mathematical Thinking and Learning* 10 (2008): 313–40.

Stein, Mary K., Jennifer L. Russell, and Margaret S. Smith. "The Role of Tools in Bridging Research and Practice in an Instructional Improvement Effort." In *Disrupting Tradition: Research and Practice Pathways in Mathematics Education*, edited by William F. Tate, Karen D. King, and Celia Rousseau Anderson, pp. 33–44. Reston, Va.: National Council of Teachers of Mathematics, 2011.

Stein, Mary K., Margaret S. Smith, Marjorie A. Henningsen, and Edward A. Silver. *Implementing Standards-Based Mathematics Instruction: A Casebook for Professional Development.* 2nd ed. New York: Teachers College Press, 2009.

Sztajn, Paola, Amy J. Hackenberg, Dorothy Y. White, and Martha Allexsaht-Snider. "Mathematics Professional Development for Elementary Teachers: Building Trust Within a School-Based Mathematics Education Community." *Teaching and Teacher Education* 23 (2007): 970–84.

Thorkildsen, Ron, and Melanie R. Scott Stein. "Fundamental Characteristics of Successful University-School Partnerships." *The School Community Journal* 6 (Fall/Winter 1996): 79–92.

CHAPTER 10

Working *with*, Supporting, and Empowering Mathematics Teaching Professionals
A Model of Effective Leadership

Michelle Cirillo
Tammie Cass
Darin Dowling
Jean Krusi
Lana Lyddon Hatten
Jeff Marks
Angie Shindelar

A S TEACHERS, WE are working during a challenging time in education. In the current climate, standards are becoming increasingly important, demands for accountability of teachers are at an all-time high, and the "intensification of teaching" (Apple 1992) seems to burden us more and more each day. Stakeholders in education agree that, in order to meet these challenges, teachers need to take part in sustained professional development; unfortunately, much of the research indicates that professional development, as typically carried out, is ineffective (Hawley and Valli 1999). With few exceptions, we feel that in our own experiences with professional development we have not always been treated as professionals, and we often did not grow in our development as mathematics teachers. This is mostly a result of what we consider "typical" professional development, which consists of short-term, ever-changing initiatives that are not sustained over substantial periods of time. In fact, one of us experienced at least seventeen different district-wide initiatives over the course of eighteen years in one district. Such hodgepodge, "hit-and-run" professional development (Loucks-Horsley et al. 2010, p. 4) has been criticized for being decontextualized and contrived (Wilson and Berne 1999).

Fortunately, evidence shows that professional development is now becoming more purposeful and intentionally designed (Loucks-Horsley et al. 2010). For example, current literature on both professional development and professional learning communities indicates that teachers need to be viewed as partners in the professional development process (Ball 1996; Stoll et al. 2006). In fact, as Weissglass (1994) pointed out in the NCTM Yearbook on professional development:

> There are two unavoidable realities about educational change: (1) The most important aspects of how the classroom operates are under the control of the classroom teacher, and (2) teachers have feelings about what they are doing and what they are asked to do. (p. 78)

Because we believe that teachers' feelings and voices *must* be considered in any successful professional development experience, we use our voices in this chapter to speak to school leaders, teacher educators, and university researchers. We hope that our experiences within an empowering professional development project and our feelings about it will encourage other professional development leaders to consider their critical roles in leading other professionals who are working to develop in their teaching of mathematics.

We are six teachers and a teacher educator who all have experience teaching middle-grades mathematics in public schools: Darin, Jeff, and Lana in urban schools, Jean in a suburban school, and Tammie and Angie in a rural school. Michelle, a former secondary mathematics teacher, was a graduate research assistant at the time that we all worked together on the project that we describe in the next section. We wrote this paper without the project leader (a teacher educator and former middle-grades teacher) because we felt that it was important to discuss and describe some of the aspects of the leadership that made the experience so meaningful and even transformative to our practices and perspectives.

The group described in this chapter was like a professional learning community (PLC) in that we were a group of professionals coming together to learn in a supportive community. As in a PLC, we had a shared leadership, supportive conditions, and shared personal practices (Hord 1997). Our group does not fit the common description of a PLC, however, in that we came together from different sites rather than from a single school. Also, we did not necessarily share a common vision, another usual feature of PLCs (Hord 1997). Instead, we shared a common desire to work on improving our classroom discourse, and we used an action research model to guide us through that process.

At professional conferences, we have stated that a key element in the success of the project was its strong and purposeful leadership. Here, we write about the Discourse Project as a model of effective leadership from the perspective that any model of effective professional development is intimately tied to its leader or leaders. We attempt to unpack some of the important ways that the leader supported our growth as learners and teachers. We do this for two reasons: First, we hope that those who lead teachers in professional development will benefit from this discussion, and second, we find it imperative that the voices of teachers be included in this important volume on professional collaborations. We have been empowered to believe that teachers' voices are no less important than those of the university researchers and school administrators who typically author chapters in volumes such as this.

After providing brief descriptions of context and methodology (i.e., the project, its participants, the kinds of things it taught us, and the research methods used for this paper), the remainder of the chapter focuses on three key elements that supported our learning. These elements helped to establish a culture within our group that was different from our prior experiences with professional development and was productive to our growth as teachers of mathematics. First, we describe the leader's approach to this project as one of working *with* us as professional collaborators, rather than conducting research *about* us or viewing professional development as something to be done *to* us. Next, we discuss the ways that we felt supported throughout the project, both pragmatically and emotionally. Finally, we describe how we felt empowered throughout the project and as a result of it.

The People and the Project

A central goal of the NSF-funded Discourse Project was to learn if and how teachers' attention to their classroom discourse might impact their beliefs and practice over time. The project began in August 2004 and continued for five years. During the first year of the project, eight middle-grades teachers were recruited from seven different schools in the Midwestern region of the United States. Teachers were purposefully selected to diversify gender, school settings, teaching experience, type of curriculum materials used, and so forth. At the end of that first year, we met for a weekend retreat to get to know one another and learn more about the work that we would participate in over the next several years. The project then progressed through several phases.

During the second year of the project, baseline data were collected in the form of classroom observations, teacher interviews, and beliefs surveys. Over the next summer and the years that followed, we met regularly as a group. Together, we read professional literature about classroom discourse and action research, reflected on our video data, and planned and discussed our action research projects. After identifying our "performance gaps"—discrepancies between what we wanted our practice to look like and what it was in reality (Hopkins 2002)—we conducted cycles of action research to work on some aspect of our practice that we wished to improve. Over the course of the five years that we worked together, we learned a great deal. (See Herbel-Eisenmann and Cirillo [2009] for elaboration on what the eight teachers learned from their participation in the Discourse Project.) In the next section, we attempt to briefly capture the kinds of things that we learned before returning to our primary focus on the effective leadership model that made this learning possible.

Developing a Discourse about Discourse

Over the course of the project, we spent a good amount of time reading and discussing literature on mathematics classroom discourse. Not only did we learn about NCTM's Discourse Standards (NCTM 1991), but we also began to think deeply about how to enact them. To this end, we discussed readings about various aspects of classroom discourse, such as classroom interaction patterns, questioning, and the use of vague language and pronouns. We read about discourse that is particularly mathematical (for example, justification and argumentation). More specifically, through the lens of discourse, we examined videos of our teaching and discussed our own discourse patterns. That is, the information we gathered in the readings helped us to develop a language to talk about our classroom talk. This is important because language, and in particular, *naming* objects or ideas, plays an important role in our learning and our thinking (Bochicchio et al. 2009; Vygotsky 1978).

Having words or phrases to name particular aspects of our practices allowed us to talk about them more thoughtfully and target them more explicitly. For example, after learning about *revoicing* (which refers here to any form of restating an idea presented by another; O'Connor and Michaels [1993]), Jean started questioning whether or not the ways in which she revoiced her students' contributions were productive toward meeting her discourse goals. As another example, Angie decided to incorporate some of the talk moves she learned about in *Classroom Discussions* (Chapin, O'Connor, and Anderson 2003) to improve her facilitation of whole-class discussions. Because this paper is about a professional development model that helped facilitate learning, and not the project itself, we return to this focus after providing some information about how data for this paper were collected and organized.

Methods

To collect the data in this chapter, the participating teacher-researchers wrote responses to the following six prompts:

- How has this experience been similar to and different from other professional development experiences that you've participated in? Please be specific.
- What has been the most meaningful to you about this experience?
- Please say something about the makeup of the group. [The group leader] purposefully selected participants who were different from each other [examples provided]. . . . What are your thoughts about this group composition? Would you have had a different preference? What were the advantages and disadvantages to this makeup?
- What norms for participation were established in this group? How were they established?
- What have you learned about yourself and your practice through this professional collaboration?
- How do you anticipate this experience will influence you as a teacher in the future?

After the teacher-authors responded to these questions, Michelle used open coding (Esterberg 2002) to develop themes for this study. The three themes developed from the written teacher-authors' responses were the ways in which the project leader (1) worked *with*, (2) supported, and (3) empowered us as professionals as we sought to improve our practice. The written responses were then organized by these themes into the sections that follow. Each author was asked to read and provide feedback on the full text, and (with only a few exceptions as noted) we write in one collective voice because the opinions expressed below were felt by all of the authors.

Our goal here is to describe the culture of the professional development that we participated in under the guidance of an effective leader. As will be demonstrated, the three themes found in our data are consistent with descriptions of effective professional development found in the literature.

Working *with* Teachers as Professional Collaborators

Prior to participating in the Discourse Project, we typically had not been held accountable for implementing changes to our practices as teachers. Nor were we provided with a rationale for change. In fact, both rationale and vision seemed to be missing from most of our past experiences. Even more important, we were rarely, if ever, provided with choices about our own development or asked to define any of the experiences for ourselves. The nature, tone, and leadership of the Discourse Project were vastly different from our former experiences in that over a five-year time period we were treated as professional collaborators, and not as teachers who needed to be told by more-knowledgeable others what and how we should change.

Enacting a Vision of Collaboration

Throughout our collaboration on the Discourse Project, we were impressed by the respect we were shown. We did not initially expect to be treated as genuine collaborators with thoughts and actions valuable enough to shape the progression of the project work. Though the teachers and researchers had specific and differing goals and roles, we always felt that we were a group working *together* to figure out how to improve mathematics classroom discourse. More specifically, the university researchers took responsibility for suggesting readings to choose from, booking meeting rooms, ordering food, and sending out reminders about the dates, times, and locations of the meetings. The teacher-researchers eventually took over the role of data collection by video-recording class sessions, collecting student work, and incorporating formative assessment. Norms and (eventually) meeting agendas were co-constructed through input from all group members.

Rather than treating us as participants in a predetermined research study, the project leader provided a vision of the project as well as support for the work undertaken. We established and revised our norms as our work evolved. For example, after expressing the need for more reflection time, we began each meeting with an hour of quiet time to review our classroom videos or student work. Because she was very tuned in to our needs, the project leader made us feel as if the project was as much ours as it was hers. And, although our work was supported by effective leadership, the decision about *what* to improve and *how* to proceed ultimately rested with each individual teacher. So while we were treated more professionally than in the typical professional development setting, we were also held more accountable.

Developing a Sense of Accountability

We felt the project placed a high value on the commodity of time. Our time was valued as far as when the meetings began and ended. Every effort was made to accommodate busy schedules and lives. Every detail of the meetings seemed to be well thought out and carefully planned to meet the needs of the group. As we became more involved in the project, we began to feel accountable to the leader as well as to each other. Lana noted, "We were shown so much respect, and I respected the leader and other participants so much that I really had a lot of trust to try to think carefully about just about anything that came up."

High expectations for discussion and task completion also contributed to the success of this experience. For example, when it was time to discuss the readings that *we* had selected, we felt that, as members of the group, we were responsible for engaging in the discussions. We were treated as if we had valuable perspectives, and we were expected to share them. The readings offered strategies for equitable participation that we enacted. Inclusivity was modeled at times subtly, explicitly, or through humor. Participation during group meetings was strongly encouraged and even specifically requested. For example, Darin recalled being "called on" to participate (e.g., "Hey, Darin, we haven't heard from you yet"). Providing him with wait time and, if necessary, offering to come back to him later helped Darin formulate and articulate a contribution. As teachers, we try to include all of our students, so we felt that it was only appropriate that the same practice occur within our group.

Many of the descriptions provided here about how the project leader worked with us as professional collaborators are consistent with contemporary beliefs about effective professional development found in the research literature. For example, Ball (1996) argued that teacher educators should model the approaches that they are promoting, and she claimed that teacher development is especially productive when teachers are in charge of the agenda and have input into the focus of the programming offered.

Supporting Our Work

In addition to being professional collaborators within the Discourse Project, we were supported in many ways—both pragmatic and emotional—to accomplish our individual goals.

Pragmatic Support

We felt that the project leader was selective about what she prepared and gave us to read and discuss during the time we spent together. The articles and books we read were continually referenced and used during the entire experience. This process helped us establish a rationale for our work. We found the reading and discussion components to be very meaningful. In particular, the practitioner and research articles focused on mathematics classroom practice helped us to think about our own practice, provided a discourse to discuss what was happening in our classrooms, and gave us tools to make the changes that we determined to be necessary. We appreciated that we were provided with the original research, afforded the time to discuss and reflect on it, and considered capable of engaging with the ideas. This approach was different from past experiences where research was either summarized for us or we were simply told to read it in our "free" time. And although readings and books were suggested and gathered for us, we were given choices about which articles to read. The leader's role during discussions could best be described as facilitator. She invited us to share, pushed our thinking by asking questions about our observations, and usually did not give answers. Just as we try to do with our own students, she posed questions that left us thinking further than we might have done on our own. The discussions focused on the readings, and our teaching practices provided us with pragmatic support to help us work on our collective goal of improving the classroom discourse.

Emotional Support

Many professional development efforts in mathematics education tend to focus on "teachers' knowledge, understandings, or behavior, rarely addressing the emotional aspects of our lives" (Weissglass 1994, p. 69). Weissglass argued, though, that the high levels of stress experienced by both teachers and administrators affect our ability to even consider, much less sustain, the process of change. As do all teachers, in addition to our stressful lives at school we also have lives outside of it. During the course of the project, some of us lost family members, and some welcomed new members into our families. Some of us sent our kids away to college, while others parented children through challenging circumstances. The members of our community often shared such events with one another. When needed, brief absences from the project work were supported and, upon return, everyone was welcomed back. This kind of sharing acknowledged our emotional needs and helped us to grow closer together as a community. This closeness facilitated a productive trust within the group that we rarely, if ever, establish with our other colleagues.

One critical component that met our physical and emotional needs was the provided meal. The meal, which was always part of the group meetings, was usually not elaborate, but rather a simple boxed lunch or pizza. Mealtime was not only one more way to help us feel valued (after all, business professionals usually expect to be fed during their longer meetings), but it was also an important time to get to know one another and share our lives both in and out of the classroom. We shared stories and created our own stories as a group. Inside jokes about ourselves and the project work became common, and we became comfortable laughing with and teasing one another. This level of comfort helped us feel safe when sharing our professional challenges and triumphs as well.

After the meal, we often shared successes, frustrations, and questions about our action research, and some of us even began sharing video clips of the attempts we were making to improve our classroom discourse. This sharing of our practice eventually became somewhat expected, though we were always invited and encouraged rather than told to do so.

Because of our emotional connection to the group members, we all felt obligated to do our part. The leader seemed to realize that for the group to be successful, she had to help us feel connected to one another and accountable to the group. As Jeff noted, it is harder to let down friends than ordinary colleagues with whom many of us do not feel a bond. We should make clear that although we have mentioned our "other" colleagues several times, we do not intend to disparage them. We only point out that, unfortunately, we have not been given the opportunity to do the kind of work and bonding described here within our home districts.

Finding Support Off-Site

Much educational research (including that related to professional learning communities) recommends that for professional development to be effective, it must be school-based (e.g., Hawley and Valli 1999). While we understand the merits of this argument for systemic change, we have found little evidence of a capacity for building such support in our home districts. For this reason, we argue that a professional development community formed outside a school setting can have a positive effect on teacher practice and student learning.

First, we appreciated that we were able to collaborate with like-minded professionals. We all chose to participate in the Discourse Project, as opposed to being mandated to do so. We also appreciated that we were all middle-grades *mathematics* teachers. Our past professional development experiences were often generalist sessions, and not focused on the particular challenges of teaching and learning mathematics. In this project, we were able to focus on mathematics learning and improving our classroom discourse.

We often noted throughout the experience that we would not have had the same level of comfort with colleagues in our own departments. There is something to be said for coming from different buildings and getting away from the local factors (e.g., petty rivalries or discussions about particular students or courses) present within our schools. In fact, two of us are from small, rural districts, where we alone were the mathematics teachers for our grades. Participation in this project helped combat the isolation we have often felt in these settings. We were afforded the luxury of stepping outside of our teaching day and into a different climate of professional collaboration.

Two of us did find some disadvantages in the fact that the collaboration was not schoolwide. For example, Lana often wished that her entire department was working on the same ideas so that she could collaborate with them locally. This sentiment gives credence to the importance of professional learning communities where the professional development is site-based and includes administrators and coaches. Overall, however, we found many advantages to the group composition of the Discourse Project.

The Discourse Project provided us with pragmatic and emotional support outside of a school setting that was both unexpected and instrumental in our professional growth.

We, like 90 percent of teachers in the United States (Darling-Hammond et al. 2009), were used to short-term conferences or workshops that were often disconnected from our practices. Research suggests that effective professional development is intensive, ongoing, focuses on the teaching and learning of specific academic content, and builds on strong working relationships among teachers (Darling-Hammond et al. 2009). Receiving this type of support within our content area in a supportive community for five years helped us to commit to and welcome opportunities to discuss and improve our discourse practices.

Professional Empowerment

The treatment and support described above empowered us to believe we are capable of making important instructional decisions, and it helped us to believe we have something important to contribute to the teaching and research communities. The regional and national presentations (at both practitioner and research conferences) and professional writings that we have done about this work have contributed to our learning and stretched us both personally and professionally.

Action research, which has been described in the literature as empowering, provided us with an unfamiliar autonomy that was crucial to our development as teaching professionals. The autonomous participation supported through the project was one of the biggest differences about the experience. As Wilson and Berne (1999) noted, learning through the type of *critical colleagueship* (Lord 1994) that was practiced in the Discourse Project is not easy:

> Perhaps the most formidable challenge is one endemic to all education. Learning, real learning, is hard work. You read, you think, you talk. You get something wrong, you don't understand something, you try it again. Sometimes you hit a wall in your thinking, sometimes it is just too frustrating. (p. 200)

To help us deal with the challenges and frustrations of this intense learning, the project leader visited us individually and provided us with the time and space to share our dilemmas and work through them with the group. We were required to choose what part of our practice to attend to and how to attend to it. We were continually reminded that we were the experts of our classrooms, and that we were the ones best able to see what and how to improve because we knew our school climate, our students, our curriculum, and our own beliefs about teaching and learning.

Because they were significant and complex, the change to classroom discourse practices we sought seemed to progress more slowly than some other things we might have done to improve our teaching. Despite the project coming to a formal end, we all feel we are engaged in ongoing efforts to improve our classroom discourse, and we have direction for individual growth. We now desire to understand the classroom more deeply and accurately from our students' perspectives. We wish to encourage *all* of our students to think more deeply about mathematics and to make their reasoning more explicit.

Finally, the Discourse Project continues to empower us in other ways. Since our work together ended, Tammie completed a master's degree program. Angie took a math coaching position. The rest of us advocated for and implemented new curricula that provided us with high-level mathematical tasks and helped us better align our practices with our beliefs. All of us have gained the confidence necessary to try out new ideas and strategies in our classrooms. This experience ratcheted up our level of professionalism and our seriousness about our important roles as facilitators of the mathematics discourse in our classrooms.

Typically, U.S. teachers have limited influence in critical areas of school decision making. In many other industrialized nations, including those that have been recognized as high achieving on international measures, teachers are actively involved in decision-making processes, and they guide much of the professional development they experience (Darling-Hammond et al. 2009). Our participation in the Discourse Project helped us to realize that professional development can be empowering and even transformative when teachers' voices and perspectives are seriously considered.

Concluding Thoughts

It is important to acknowledge how fortunate we are to have been supported, not only by an effective leader, but also by NSF funds. We understand that not all teachers, administrators, and districts have the luxury of development funds. However, there are many ways that the spirit of what was described here can be carried out locally. For example, treating teachers as professional collaborators is free. Supporting teachers emotionally is free. And helping teachers feel empowered by providing them with choices about what and how to improve is also free.

We stated earlier that a goal of this project was to learn if and how attending to our classroom discourse might impact our beliefs and practices over time. What we have learned is that attending to our discourse was less about *changing* our beliefs, and more about strengthening our professed beliefs and better aligning those beliefs with our practices. We were able to reach this conclusion because we were supported to try to be better teachers in a manner that modeled how we want to be better teachers. Our leader helped us know that even small changes can take a long time, but the impact of thoughtful small changes can be worth the wait. We were expected to decide which small changes to try and given time to try them. The most meaningful thing to us about this experience was that we were invited, given rationale, and then helped to carve out the time, space, and energy to make specific changes in our practices so that our students could have the kinds of experiences that we believe are important for them.

......

The authors would like to thank the volume editor and editorial panel whose feedback helped improve the quality of this chapter. We would also like to acknowledge Joe Obrycki and Patty Gronewold, the two other teacher-researchers who contributed to this meaningful experience. This research was supported, in part, by the National Science Foundation under grant number 0347906 (Beth Herbel-Eisenmann, Principal Investigator). Any opinions, findings, and conclusions or recommendations expressed in this material are those of the authors and do not necessarily reflect the views of the National Science Foundation.

REFERENCES

Apple, Michael W. "Do the Standards Go Far Enough? Power, Policy, and Practice in Mathematics Education." *Journal for Research in Mathematics Education* 23, no. 5 (1992): 412–31.

Ball, Deborah L. "Teacher Learning and the Mathematics Reforms: What Do We Think We Know and What Do We Need to Learn?" *Phi Delta Kappan* 77 (1996): 500–508.

Bochicchio, Daniel, Shelbi Cole, Deborah Ostien, Vanessa Rodriguez, Megan Staples, Patricia Susla, and Mary Truxaw. "Shared Language." *Mathematics Teacher* 102, no. 8 (2009): 606–13.

Chapin, Suzanne H., Catherine O'Connor, and Nancy Canavan Anderson. *Classroom Discussions: Using Math Talk to Help Students Learn.* Sausalito, Calif.: Math Solutions Publications, 2003.

Darling-Hammond, Linda, Ruth Chung Wei, Alethea Andree, Nikole Richardson, and Stelios Orphanos. *Professional Learning in the Learning Profession: A Status Report on Teacher Development in the U.S. and Abroad.* Oxford, Ohio: National Staff Development Council, 2009.

Esterberg, Kristin G. *Qualitative Research Methods.* New York: McGraw-Hill, 2002.

Hawley, Willis D., and Linda Valli. "The Essentials of Effective Professional Development." In *Teaching as the Learning Profession: Handbook of Policy and Practice,* edited by Linda Darling-Hammond and Gary Sykes, pp. 127–50. San Francisco: Wiley & Sons, Inc., 1999.

Herbel-Eisenmann, Beth, and Michelle Cirillo, eds. *Promoting Purposeful Discourse.* Reston, Va.: NCTM, 2009.

Hopkins, David. *A Teacher's Guide to Classroom Research.* Buckingham, UK: Open University Press, 2002.

Hord, Shirley M. *Professional Learning Communities: Communities of Continuous Inquiry and Improvement.* Austin, Tex.: Southwest Educational Development Laboratory, 1997.

Lord, Brian. "Teachers' Professional Development: Critical Colleagueship and the Role of Professional Communities." In *The Future of Education: Perspectives on National Standards in Education,* edited by Nina Cobb, pp. 175–204. New York: College Entrance Examination Board, 1994.

Loucks-Horsley, Susan, Katherine E. Stiles, Susan Mundry, Nancy Love, and Peter W. Hewson. *Designing Professional Development for Teachers of Science and Mathematics.* 3rd ed. Thousand Oaks, Calif.: Corwin, 2010.

National Council of Teachers of Mathematics. *Professional Standards for Teaching Mathematics.* Reston, Va.: NCTM, 1991.

O'Connor, Mary Catherine, and Sarah Michaels. "Aligning Academic Task and Participation Status through Revoicing: Analysis of a Classroom Discourse Strategy." *Anthropology & Education Quarterly* 24 (1993): 318–35.

Stoll, Louise, Ray Bolam, Agnes McMahon, Mike Wallace, and Sally Thomas. "Professional Learning Communities: A Review of the Literature." *Journal of Educational Change* 7 (2006): 221–58.

Vygotsky, Lev S. *Mind in Society: The Development of Higher Psychological Processes.* Cambridge, Mass.: Harvard University Press, 1978.

Weissglass, Julian. "Changing Mathematics Teaching Means Changing Ourselves: Implications for Professional Development." In *Professional Development for Teachers of Mathematics*, 1994 Yearbook of the National Council of Teachers of Mathematics (NCTM), edited by Douglas B. Aichele and Arthur F. Coxford, pp. 67–78. Reston, Va.: NCTM, 1994.

Wilson, Suzanne M., and Jennifer Berne. "Teacher Learning and the Acquisition of Professional Knowledge: An Examination of Research on Contemporary Professional Development." *Review of Research in Education* 24 (1999): 173–209.

Part IV
Teacher Leaders and Coaches

How can we develop teachers into leaders within and across schools?

One of the greatest challenges facing educators today is how to align instructional practices in ways that improve student learning and create equity. This challenge has arisen from the creation of more rigorous standards for student mathematical learning. With these new expectations for learning has come the need for more rigorous standards for teaching.

One solution to the conundrum of improving and aligning teacher practice is to use coaches as change agents. Yet coaches—even when they are effective (and, as documented in Patricia F. Campbell's research in chapter 11, not all models of coaching or uses of coaches' time are equally effective in supporting teacher development)—are still confronted by the enormity of the task. Another possible solution is to have professional learning communities that focus on the instructional core (planning, co-teaching, and debriefing lessons).

The last two chapters in Part IV highlight two examples of teacher leadership. In each chapter, teachers' leadership potential was cultivated in two key ways: first, through on-site in-classroom coaching; and, second, through teachers' ongoing experiences in professional learning communities designed to deepen pedagogical and content knowledge and to create collaborative teacher networks.

Questions for this section:

- How does coaching impact teacher beliefs and dispositions, knowledge and competencies, and ultimately affect teaching and learning?
- What are the characteristics of teacher leadership, and how might professional learning communities develop this capacity in teachers?
- What is the most effective use of a coach's time? What evidence do you have to support your thinking?
- What rationale can be offered for collaborations designed to explore the instructional core (planning, co-teaching, and debriefing lessons)?
- How might coaching or other teacher leadership models be focused to reinvent teacher practice at scale?

CHAPTER 11

Coaching and Elementary Mathematics Specialists

Findings from Research

Patricia F. Campbell

R ESEARCH STUDIES HAVE SHOWN that the quality of delivered instruction is a critical factor influencing student achievement (Rivkin, Hanushek, and Kain 2005). Because sporadic teacher workshops addressing unrelated topics do not yield sustained professional growth (Knapp 2003), school districts across the nation are turning to elementary mathematics specialists or coaches to lead the way to improved achievement. This policy is based on the conviction that schools need to become places where teachers can learn (Hawley and Valli 1999). The intent is for a respected, knowledgeable colleague—one who couples instructional expertise with a deep understanding of mathematics and students— to serve as an available, on-site resource for teachers. In many schools, the elementary mathematics coach is expected not only to work alongside teachers in their classrooms, fostering enhanced teaching practices, but also to increase a school's instructional capacity (Neufeld and Roper 2003). In such a system, the coach is often charged with serving as a school's on-site specialist for mathematics; catalyzing and sustaining efforts that span mathematics curriculum, instruction, and assessment; and encouraging collective professional habits that will advance schoolwide change as well as student learning and achievement (Campbell and White 1997; York-Barr and Duke 2004). A specialist/coach's work may be defined solely as working with individual teachers, or it may also support change across a school's mathematics program, but either way it quickly becomes intertwined with establishing and supporting collaborative relationships.

Many of the policies that led to the creation of elementary mathematics coach or specialist positions were made in response to pressure to raise student achievement. It is only now that we are seeing reports that address the complexity of the position as well as its effectiveness. This chapter presents a review of that work, first considering the intended roles for coaches as well as the challenges they encounter. This is followed by a report on the findings and implications of a study designed to investigate the impact of elementary mathematics specialists on student achievement and teacher beliefs in grades 3–5.

Multiple Roles

As research identified sustained and job-embedded professional development as more likely to be associated with changed instructional practice and increased student learning (Darling-Hammond and Richardson 2009), school districts began to place coaches in schools. Figure 11.1 shows a model that builds on Desimone's work (2009), displaying how the role of elementary mathematics specialist and coach may interact with other forms of professional development within the complex setting of a school.

Some school-based professional development designs focus solely on content-focused coaching. These efforts target individual teachers and grade-level teams via strategies such as co-planning, co-teaching, observation, demonstration teaching, debriefing, and mentoring. While content-focused coaching encompasses important aspects of their responsibilities, elementary mathematics specialists may also be called on to provide programmatic leadership by assuming the role of "community organizer" for mathematics in their schools (Neufeld and Roper 2003). For example, a statewide mathematics specialist effort in Virginia did specify "collaborate with individual teachers through co-planning, co-teaching, and coaching" as an expectation for school-based work, but it also asked specialists to—

- assist administrative and instructional staff in interpreting data and designing approaches to improve student achievement and instruction;
- ensure that the school curriculum is aligned with state and national standards and their school division's mathematics curriculum;
- promote teachers' delivery and understanding of the school curriculum through collaborative long-range and short-range planning;
- facilitate teachers' use of successful, research-based instructional strategies, including differentiated instruction for diverse learners such as those with limited English proficiency or disabilities;
- work with parents/guardians and community leaders to foster continuing home/school/community partnerships focused on students' learning of mathematics; and
- collaborate with administrators to provide leadership and vision for a schoolwide mathematics program (Virginia Mathematics and Science Coalition 2008, p. 1).

These programmatic efforts, which are supplemental to content-focused coaching, further define how specialists may collaborate with parents, teachers, and administrators. Specialists may also influence how much teachers access other avenues for professional development. As depicted in figure 11.1, each of these components, along with mathematics coaching targeted to individual teachers or grade-level teams, may then have an impact on teacher knowledge, competencies, beliefs, and dispositions, and thereby

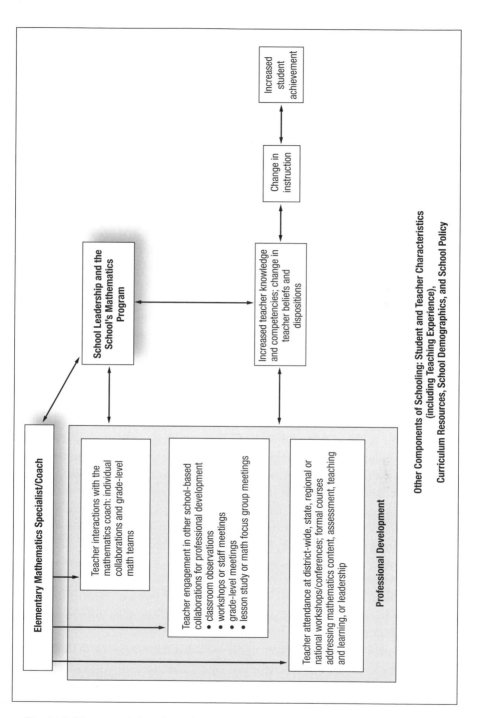

Fig. 11.1. How specialists/coaches influence professional development, classroom practice, and student learning

potentially affect instruction and achievement. Other elements also influence instruction, such as teachers' attention to and management of students; students' interpretation of and engagement in instructional tasks; the quality of available resources; the intended curriculum; teaching experience; and the nature of student-teacher interactions, as well as a myriad of situational factors in the classroom, school, and district.

Transitions and Challenges

Elementary mathematics specialists and coaches are often selected from the ranks of accomplished teachers (Chval et al. 2010). Published reports have detailed the challenges that whole-school coaches or specialists encounter, as effective teachers of children must develop additional, distinct abilities in order to be effective coaches of other teachers. Some of these abilities are knowledge-based, such as understanding the mathematics curriculum across grades or courses, characterizing instructional models and change, carrying out classroom observations, interpreting data, and organizing instructional materials (Neufeld and Roper 2003). Other demands are directly associated with the shift in identity required when assuming the responsibilities of an elementary mathematics coach.

As noted by Chval and her colleagues (2010), a critical challenge that instructional coaches face is the shift from being viewed by others and by oneself as an expert (as an expert teacher) to being viewed as a novice (a novice specialist or coach). This shift from expert to novice can be difficult and disconcerting. A coach is expected to carry out on-site professional support for individual teachers; this includes not only demonstration teaching and co-teaching, but also co-planning, debriefing, mentoring, and formative feedback (West and Staub 2003). Generally new coaches gladly accept the opportunity to work collegially in classrooms with welcoming teachers, but they can also be surprised and frustrated when faced with teacher resistance or a teacher's unprofessional behavior, or when viewed as the "resident mathematics expert in their building" (Chval et al. 2010, p. 201). Coaches typically know that they will be leading grade-level mathematics planning meetings in order to catalyze teacher learning and instructional change through the strategic use of materials supporting student performance. But coaches must also learn how to support teachers while questioning them; how to frame a common goal across differing instructional philosophies while trying to build community within and across grade-level teams; how to facilitate positive discussion advancing mathematical knowledge while addressing teachers' limited understandings; and how to navigate the organizational and cultural factors that exist in schools (Bean et al. 2010; Poglinco and Bach 2004; West et al. 2007).

Because of their prior positions as highly regarded teachers of mathematics, many new elementary mathematics specialists anticipate that they will be asked by their principals to assume an instructional leadership role for mathematics, such as offering suggestions for school improvement plans. But a specialist may not feel prepared to interpret the instructional implications of schoolwide student achievement data for mathematics or to develop the entire school's capacity for district-wide change efforts. At the same time, because they have only just left the role of classroom teacher, new elementary mathematics specialists/coaches are typically unsure of how to negotiate rather than simply accept other duties that principals might want to assign, such as serving as a school's testing coordinator.

Chval and her colleagues (2010) found that new elementary mathematics coaches experience emotional responses when confronting their changing role and new identity. They are frequently disappointed in themselves for not accomplishing as much as they had expected. They are unsure of how to balance multiple responsibilities and how to set priorities within time constraints. Coaches often feel a sense of displacement, going from a position of respect in their former school to the isolation of being viewed as a nebulous addition to the professional staff. They may even feel guilty as they identify their new position with abandoning students who could have benefited from their classroom instruction. The implication of these findings is that, whether assigned as a coach or as an elementary mathematics specialist with both coaching and programmatic responsibilities, it is critical that prospective specialists/coaches, district administrators, and school principals negotiate roles and responsibilities prior to placement, and also revisit these expectations periodically (Chval et al. 2010). Further, given the demanding and encompassing nature of their position, elementary mathematics specialists and coaches require both substantive professional enhancement/coursework prior to assuming their new role and a supportive structure for professional development afterward. Only then can the intended outcomes of high-quality instructional practice and increased student achievement, as depicted in figure 11.1, be realized.

The Impact of Elementary Mathematics Specialists

To investigate the impact of elementary mathematics specialists, the National Science Foundation funded a collaborative research project involving four universities and five school districts in Virginia within a three-year, randomized, control-treatment design. This study's twenty-four treatment and twelve control schools represented a range of demographic and economic settings and included urban, suburban, and urban-edge schools. The two cohorts of twelve specialists each in this study were experienced classroom teachers who were selected by their school district and assigned to provide full-time support in a school. To prepare for this role, the specialists completed five mathematics content courses and two leadership-coaching courses that included study of models, resources, and best practices for mathematics instruction. The intent of this project was to determine the effect of knowledgeable elementary mathematics specialists in two domains. A primary focus was to study the impact of mathematics specialists on student achievement (grades 3–5), as measured by the high-stakes standardized assessment required in Virginia by the federal regulations for No Child Left Behind. Second, the project considered whether the specialists influenced the teachers' beliefs about mathematics teaching and learning.

Activity of Elementary Mathematics Specialists

These elementary mathematics specialists logged the duration and category of their daily activities using a data collection and transmittal program operating on a personal digital assistant (PDA). While the first cohort of twelve specialists did this for three years, the second cohort recorded only one year of data. These logs indicated that although specialists were not expected to complete tasks related to their work responsibilities outside of their contract day (median contract day 7.5 hours), many specialists did so. Indeed, on average, these specialists spent more than four hours per week on work-related tasks, including coaching, for which they received no financial compensation.

Table 11.1

Mean hours per contracted week of specialist activities by cohort and year

Activity	Cohort 1 2005–06 (Year 1)	Cohort 1 2006–07 (Year 2)	Cohort 1 2007–08 (Year 3)	Cohort 2 2007–08 (Year 1)
Coaching Teachers (Individual Teachers and Grade-level Teams)	8.21	4.91	4.84	3.83
Preparing for Teaching/Coaching	4.43	4.65	4.69	4.43
Supporting Assessment	3.98	5.06	5.14	4.69
Teaching or Supporting Students (Not Demonstration or Co-teaching)	1.13	1.65	1.69	1.35
Supporting the School Mathematics Program	1.88	1.58	1.91	1.91
Performing School-Based Duties	2.44	3.45	3.90	3.68
Materials Management/ Communication Tasks	3.64	4.13	4.43	4.28
Attending Meetings	3.45	2.55	2.51	3.56
Engaging in Personal Professional Activity	4.95	5.51	4.09	5.40
Noneducational Activities (lunch, travel, school events)	3.38	4.05	4.24	4.43

Table 11.1 displays the average number of hours that the specialists in each cohort spent on varying activities in a typical 37.5-hour contract week. For the most part, these specialists spent minimal time in grade-level mathematics planning sessions; their coaching activity emphasized working with individual teachers. The amount of contract-day time that the Cohort 1 specialists spent coaching individual teachers and grade-level teams decreased over the three years. When comparing the two cohorts, the time spent on coaching was more consistent when the year of work was constant (both cohorts in 2007–08) than when the extent of experience as a specialist was constant (Cohort 2 in 2007–08 and Cohort 1 in 2005–06). This may mean that during 2007–08 there were common outside influences impacting the amount of time that specialists felt they had to spend working with individual teachers and grade-level planning teams.

Over that same time period, the amount of time that Cohort 1 specialists spent addressing assessment increased. The Cohort 2 specialists spent somewhat less time than the more experienced Cohort 1 specialists addressing assessment during 2007–08, primarily because Cohort 2 specialists had less time devoted to developing assessments and to assessment management. This may reflect the increased managerial expertise presumed of Cohort 1 specialists during their third year of placement. The increase in Cohort 1 assessment activity and the substantial amount of specialist time devoted to assessment responsibilities across the two cohorts were evident in each of the five school districts, with assessment-related tasks consistently being the most common activity of the specialists in the urban districts. Because the cooperating districts did not position mathematics specialists across each of their elementary schools, the shifting of specialists' time to assessment responsibilities was probably a local school response to district pressures, and not a district-level assignment.

The time specialists spent in meetings that did have a mathematics focus was quite consistent within districts, while being unique across districts. This indicates that a specialist's attendance at a meeting addressing mathematics was likely not an individual decision, but reflected an expectation of either a principal or district office. As Cohort 1 specialists gained expertise, local administrators were less likely to expect their attendance at a meeting when the agenda was not related to mathematics.

Some of the prevalence of activity associated with personal professional development reflects that all specialists completed the second leadership-coaching course during their first year of placement. Further, approximately half of the specialists in each cohort completed an additional graduate course or two during their first year of placement as they completed requirements for a master's degree. However, the time spent on personal professional development also reflects that the role of elementary mathematics specialist requires learning after placement. Specialists met this need through the reading of professional literature and periodic professional development sessions in their districts.

Each of these specialists spent time establishing and maintaining communication with their colleagues, teachers, and parents. All of the participating school districts provided email access to their instructional and administrative staffs. Email communication time increased noticeably between 2005–06 and 2007–08. This is most likely a reflection of changes in culture that are not unique to these schools. In contrast, the prevalence of school-based duties is most likely a project-related artifact. The specialists advised each other to "volunteer for bus duty" as a way to build trust and entrée into their school placements, noting that this was a time when few, if any, teachers would be available to meet with a specialist. Time allocated to noneducational activities (see table 11.1) was primarily at midday and reflected a break for lunch, although periodic travel to half-day, off-site meetings also occurred.

Impact on Student Achievement

To determine whether these elementary mathematics specialists impacted student achievement, two analyses of student mathematics achievement scores on Virginia's standardized state assessment (grades 3, 4, and 5) were completed (Campbell and Malkus 2011). The Treatment versus Control analysis compared three years of mathematics achievement scores of students in the control schools to the scores of students in the schools with an elementary mathematics specialist. The student data from the schools with the specialists were either the three years of data from the Cohort 1 schools or the single year of Cohort 2 data collected during the third year of the study. A second Cohort-by-Year versus Control analysis compared three years of mathematics achievement scores of students in the control schools to scores of students in the treatment schools, keeping track of whether the scores from treatment schools were from the first, second, or third year of a specialist being placed in a school.

When all three years of data from grades 3, 4, and 5 were compared in the Treatment versus Control analysis, the mathematics achievement scores of students in the schools where a Cohort 1 elementary mathematics specialist was placed were significantly higher than the achievement scores of the students in the control schools. This was the case for all three grades. However, during their first year of placement, the Cohort 2 specialists did not significantly impact the mathematics achievement scores of the students in their schools, as compared to the scores of the students remaining in the control schools and as compared to the Cohort 1 schools.

The fact that the Cohort 1 specialists had a statistically significant impact on student achievement while the Cohort 2 specialists did not was an interesting finding. Was this because Cohort 1 specialists were more effective than Cohort 2 specialists? Or was it because Cohort 1 specialists had more time to work with teachers in their school mathematics program? To determine why Cohort 1 specialists had more impact on student achievement than Cohort 2 specialists, a second analysis was conducted. This analysis was done to see whether the difference in results was because Cohort 1 specialists were more effective or because Cohort 1 specialists had more time to work with teachers in their school mathematics program.

This second analysis found no statistically significant difference in student achievement between the control schools and the schools with specialists when the specialists were in their first year of placement. This was the case for both cohorts of specialists and for all three grades. However, student achievement in the schools with specialists was consistently greater than the achievement of students in the control schools during the second year of specialist placement (with a statistically significant difference in grades 4 and 5); this difference in student achievement either increased (grades 3 and 5) or was comparably maintained in the third year of the placement of a specialist (grade 4). The implication of this finding is that a specialist's positive effect on student achievement develops over time as a knowledgeable elementary mathematics specialist and the instructional and administrative staffs in the assigned school learn and work together. While elementary mathematics specialists did have a statistically significant positive impact on student achievement over time, there was no evidence that elementary mathematics specialists will yield increased student achievement in their first year of placement.

Impact on Teachers' Beliefs about Mathematics Teaching

Administrators place elementary mathematics specialists in schools in order to promote and support instructional change. However, research indicates that teachers' perceptions of mathematics teaching and learning interrelate with their instructional practices (Ross et al. 2004; Ross et al. 2003). Further, teachers' beliefs may limit what they notice when observing demonstration lessons, unless lesson observations are coupled with activities such as collegial discussions that encourage teachers' reflection about the instruction (Grant, Hiebert, and Wearne 1998). Elementary mathematics specialists may catalyze and foster these types of observations and discussions and, in this way, influence teachers' beliefs. Thus, change in teachers' beliefs about mathematics teaching and learning is another way of evaluating the effect of specialists.

In the Virginia study, the teachers in both the control schools and the schools with the mathematics specialists completed beliefs surveys about mathematics teaching and learning. The surveys were administered at the beginning of the first school year of specialist placement, or whenever a new teacher came to one of the participating schools, and then again each spring for up to four years (2005–09). This thirty-item survey was a variation of an instrument developed by Ross and colleagues (2003). Teachers rated each of thirty statements on a scale of 1 (strongly disagree) to 5 (strongly agree). The statements reflected perspectives about mathematics curriculum and instruction and about the needs of students and student understanding. Some of these items concerned beliefs that emphasized directed teaching and mathematical structure as a basis for curriculum (Traditional), and other items reflected a perspective emphasizing the development of students' principled knowledge and supporting student efforts to "make sense" of the mathematics (Making Sense). Table 11.2 presents sample survey items.

Table 11.2

Example items from the beliefs survey

Items Reflecting a Traditional Perspective
Learning mathematics requires a good memory because you must remember how to carry out procedures and, when solving an application problem, you have to remember which procedure to use.
The best way to teach students to solve mathematics problems is to model how to solve one kind of problem at a time.

Items Reflecting a "Making Sense" Perspective
Students can figure out how to solve many mathematics problems without being told what to do.
I don't necessarily answer students' math questions but rather let them puzzle things out for themselves.

Because the specialists in this study logged their daily activity on a PDA, there were data available to identify whether a teacher in a treatment school was highly engaged with a mathematics specialist or not. The analysis indicated that this was important information. Over time, the beliefs of teachers in the schools with an elementary mathematics specialist did not differ significantly in terms of the Making Sense perspective from those of teachers in the control schools, unless the teachers were highly engaged with the specialist. The beliefs of teachers who were highly engaged with a specialist changed significantly to be more in agreement with the Making Sense perspective. Similarly, over time, the beliefs of teachers in the schools with an elementary mathematics specialist did not differ significantly in terms of the Traditional perspective from those of teachers in the control schools, unless the teachers were highly engaged with the specialist. The beliefs of teachers who were highly engaged with a specialist changed significantly, shifting away from the Traditional perspective toward a Making Sense perspective.

Implications

The rationale for coaching focuses on job-embedded professional development that blends support for teachers with feedback to them. However, elementary mathematics specialists may also teach mathematics to students, support school-wide improvement in mathematics, and address the cross-grade translation and implementation of mathematics curriculum and assessment expectations (Association of Mathematics Teacher Educators et al. 2010). These responsibilities extend beyond mentoring teachers and require specialists to establish a sense of trust and credibility with teachers, administrators, and parents. The data from the study of Virginia elementary mathematics specialists and a recent study of Reading First coaches (Bean et al. 2010) identified similar findings: Whenever coaches/specialists spent substantial time attending to non-coaching demands, the patterns of activity seemed to be situational, reflecting the organizational and cultural factors present in the local school. Combining this information with the finding from the Virginia study that the positive impact of mathematics specialists emerged over time, one implication may be inferred. If elementary mathematics specialists are to serve as change agents for instructional improvement and teachers' professional growth, they must have the time and tendency to establish, develop, and maintain collaborative networks in their schools that are linked to the school's sense of community and collective, professional identity (Frank, Zhao, and Borman 2004).

Although on average the specialists in the study conducted in Virginia spent less time coaching than anticipated, the role that they played in their schools still had a significantly positive effect on student achievement. This suggests that the potential impact of elementary mathematics specialists on student achievement might be even greater if specialists primarily focused on coaching rather than other responsibilities that are often assigned to them. However, that would require principals to understand the role of the specialist as an agent and catalyst who establishes and maintains a safe coaching environment for instructional improvement in mathematics. Supportive principals and knowledgeable specialists should work together as instructional leaders in their schools. Together they may establish a critical relationship that impacts not only coaching, but also the effectiveness of the school's mathematics program. Yet, that collaboration must extend to encompass teachers. If teachers perceive a specialist as focused primarily on administrative tasks, then they will not consider that specialist as a person who can help them address instructional needs or problems (Bean et al. 2010). Indeed, "coaching is a relational practice whose efficacy presumably depends on the quality of the relationship that a coach is able to establish with each individual teacher" (Biancarosa, Bryk, and Dexter 2010, p. 29).

Research frequently offers contradictory results. Two recent studies of literacy coaches found they had inconsistent (Marsh et al. 2008) or no (Garet et al. 2008) effect on student achievement. The study of elementary mathematics specialists in Virginia and a study of Literacy Collaborative coaches (Biancarosa, Bryk, and Dexter 2010) both identified a significantly positive impact. One reason for this discrepancy may lie in the differing expectations for these specialists/coaches. In both of the literacy-coaching studies that did not identify a significant relationship between coaching and student achievement, the coaches participated in minimal professional development prior to placement. The elementary mathematics specialists in Virginia completed six graduate courses that were designed to foster and support their transition to the position of whole-school elementary mathematics specialist prior to placement. Similarly, the Literacy Collaborative coaches completed a full year of professional development prior to placement. This suggests that the content and pedagogical knowledge of coaches/specialists, as well as their understanding of the responsibilities and skills associated with accomplished leadership and coaching, are critical components for improved student achievement. The results described here should not be generalized to other settings where an experienced teacher is simply anointed as the elementary mathematics specialist/coach with little professional development.

......

This paper was developed with the support of the National Science Foundation, grant number ESI-0353360. The statements and findings herein reflect the opinions of the author and not necessarily those of the National Science Foundation.

REFERENCES

Association of Mathematics Teacher Educators, Association of State Supervisors of Mathematics, National Council of Supervisors of Mathematics, and National Council of Teachers of Mathematics. *The Role of Elementary Mathematics Specialists in the Teaching and Learning of Mathematics.* http://www.nctm.org/about/content.aspx?id=26069.

Bean, Rita M., Jason A. Draper, Virginia Hall, Jill Vandermolen, and Naomi Zigmond. "Coaches and Coaching in Reading First Schools: A Reality Check." *The Elementary School Journal* 111 (September 2010): 87–114.

Biancarosa, Gina, Anthony S. Bryk, and Emily R. Dexter. "Assessing the Value-Added Effects of Literacy Collaborative Professional Development on Student Learning." *The Elementary School Journal* 111 (September 2010): 7–34.

Campbell, Patricia F., and Dorothy Y. White. "Project IMPACT: Influencing and Supporting Teacher Change in Predominantly Minority Schools." In *Mathematics Teachers in Transition,* edited by Elizabeth Fennema and Barbara Scott Nelson, pp. 309–55. Mahwah, N.J.: Lawrence Erlbaum Associates, 1997.

Campbell, Patricia F., and Nathaniel N. Malkus. "The Impact of Elementary Mathematics Coaches on Student Achievement." *The Elementary School Journal* 111 (March 2011): 430–54.

Chval, Kathryn B., Fran Arbaugh, John K. Lannin, Delinda van Garderen, Liza Cummings, Anne T. Estapa, and Maryann E. Huey. "The Transition from Experienced Teacher to Mathematics Coach: Establishing a New Identity." *The Elementary School Journal* 111 (September 2010): 191–216.

Darling-Hammond, Linda, and Nikole Richardson. "Teacher Learning: What Matters?" *Educational Leadership* 66 (February 2009): 46–53.

Desimone, Laura M. "Improving Impact Studies of Teachers' Professional Development: Toward Better Conceptualizations and Measures." *Educational Researcher* 38 (April 2009): 181–99.

Frank, Kenneth A., Yong Zhao, and Kathryn Borman. "Social Capital and the Diffusion of Innovations within Organizations: The Case of Computer Technology in Schools." *Sociology of Education* 77 (April 2004): 148–71.

Garet, Michael S., Stephanie Cronen, Marian Easton, Anja Kurki, Meredith Ludwig, Wehmah Jones, Kazuaki Uekawa et al. *The Impact of Two Professional Development Interventions on Early Reading Instruction and Achievement* (NCEE 2008-4030). Washington, D.C.: National

Center for Education Evaluation and Regional Assistance, Institute of Education Sciences, U.S. Department of Education, 2008.

Grant, Theresa J., James Hiebert, and Diana Wearne. "Observing and Teaching Reform-Minded Lessons: What Do Teachers See?" *Journal of Mathematics Teacher Education* 1 (1998): 217–36.

Hawley, Willis. D., and Linda Valli. "The Essentials of Effective Professional Development: A New Consensus." In *Teaching as the Learning Profession: Handbook of Policy and Practice,* edited by Linda Darling-Hammond and Gary Sykes, pp. 127–50. San Francisco: Jossey-Bass, 1999.

Knapp, Michael S. "Professional Development as a Policy Pathway." *Review of Research in Education* 27 (January 2003): 109–57.

Marsh, Julie A., Jennifer Sloan McCombs, J. R. Lockwood, Francisco Martorell, Daniel Gershwin, Schott Naftel, Vi-Nhuan Le et al. *Supporting Literacy across the Sunshine State: A Study of Florida Middle School Reading Coaches* (MG-762-CC). Santa Monica, Calif.: RAND, 2008.

Neufeld, Barbara, and Dana Roper. *Coaching: A Strategy for Developing Instructional Capacity.* Cambridge, Mass.: Education Matters, 2003. http://annenberginstitute.org/pdf/Coaching.pdf.

Poglinco, Susan M., and Amy J. Bach. "The Heart of the Matter: Coaching as a Vehicle for Professional Development." *Phi Delta Kappan* 85 (January 2004): 398–400.

Rivkin, Steven G., Eric A. Hanushek, and John F. Kain. "Teachers, Schools, and Academic Achievement." *Econometrica* 73 (March 2005): 417–58.

Ross, John A., Douglas J. McDougall, Cathy Bruce, Sonia Ben Jafaar, and Jane Lee. *A Multi-Dimensional Approach to Mathematics In-service.* Paper presented at the annual meeting of the North American Chapter of the International Group for the Psychology of Mathematics Education, 2004. http://www.allacademic.com/meta/p117501_index.html.

Ross, John A., Douglas McDougall, Ann Hogaboam-Gray, and Ann LeSage. "A Survey Measuring Elementary Teachers' Implementation of Standards-Based Mathematics Teaching." *Journal for Research in Mathematics Education* 34 (July 2003): 344–63.

Virginia Mathematics and Science Coalition. *DR K-12 Support Materials,* 2008. http://www.vamsc.org.

West, Lucy, Ginger Hanlon, Phyllis Tam, and Milo Novelo. "Building Coaching Capacity through Lesson Study." *NCSM Journal of Mathematics Education Leadership* 9 (Winter 2007): 26–33.

West, Lucy, and Fritz C. Staub. *Content-Focused Coaching: Transforming Mathematics Lessons.* Portsmouth, N.H.: Heinemann, 2003.

York-Barr, Jennifer, and Karen Duke. "What Do We Know about Teacher Leadership? Findings from Two Decades of Scholarship." *Review of Educational Research* 74 (September 2004): 255–317.

CHAPTER 12

Using Instructional Coaching

Customized Professional Development in an Integrated High School Mathematics Program

Erin Elizabeth Krupa
Jere Confrey

PROFESSIONAL DEVELOPMENT WORKSHOPS, held during the summer for a one- or two-week period or during a teacher workday, continue to be the most common means to promote changes in teachers' instructional practices or to strengthen their content knowledge (Darling-Hammond et al. 2009; Stein et al. 1999). Yet, research has shown convincingly that teachers are not likely to change their instructional practices solely by attending isolated professional developments, and that ongoing support can help teachers implement the ideas presented in these professional developments (Ball 1996; Ball and Cohen 1999; Guskey 2002; Loucks-Horsley 2010; Putnam and Borko 1997; Wilson and Berne 1999). As summarized succinctly in *Principles and Standards for School Mathematics* (NCTM 2000):

> The reality is simple: unless teachers are able to take part in ongoing, sustained professional development, they will be handicapped in providing high-quality mathematics education. The current practice of offering occasional workshops and in-service days does not and will not suffice. (p. 369)

Less well documented are the ways to provide continuous support for the types of collaborations necessary to facilitate improvements in teacher practices. This chapter reports on one attempt to create a sustainable professional development, centered on the use of an instructional coach who provides customized support for teachers during the year as an extension of a summer program.

Instructional coaches, as envisioned in *Principles and Standards for School Mathematics* (NCTM 2000), are hired by school districts as mathematics teacher-leaders who help teachers continually improve their mathematics instruction. The coaches facilitate interactions among teachers and provide them with the customized support they need to make changes to their classroom practices. Ball (1996) states, "The most effective professional development model is thought to involve follow-up activities, usually in the form of long-term support, coaching in teachers' classrooms, or ongoing interactions with colleagues" (pp. 501–2). Through an analysis of research on different types of teacher training, Joyce and Showers (2002) found that training programs that incorporate a coaching component are more effective in helping teachers transfer knowledge and skills into practice than trainings without coaching. In the remainder of this chapter, we report on a case study of the use of instructional coaches in an integrated mathematics professional development project.

The North Carolina Integrated Mathematics Project

The North Carolina Integrated Mathematics Project (NCIM) was developed to create and support a community of teachers using the *Core-Plus Mathematics* (CPMP) integrated curriculum materials, particularly in high-needs schools. Its aim was to educate teachers about the content and pedagogy of using integrated mathematics by creating a sustainable professional development model that could be replicated across the nation. Located in various rural parts of the state, the seven partner schools were identified as low performing (based on North Carolina accountability measures). The student population for the project schools for the 2008–09 school year ranged between 110 and 163 students, and the average ethnic makeup consisted of 1 percent American Indian or Asian, 6 percent Hispanic, 16 percent white, and 77 percent black. On average, 72.8 percent of students at each school qualified for free and reduced lunch. The schools were part of a collaboration called the New Schools Project, and each had a focus on STEM (science, technology, engineering, and mathematics) education. Twelve teachers were actively involved in the project, some who were the only mathematics teacher at their school. One of the project's goals was to establish statewide collaborations to assist rural schools with the challenges of isolation.

Four Components of the Professional Development Model

The coaching model was situated within the larger project of preparing teachers to implement *Core-Plus Mathematics*, all in order to strengthen and invigorate STEM education at these schools. The overall professional development model was an attempt to improve and strengthen teachers' mathematical content and pedagogical knowledge. Working together, the four main elements of the professional development model included a summer workshop providing in-depth education on the use of curricular materials (one or two weeks); a Web-based environment supporting information exchange; face-to-face follow-up conferences; and monthly instructional coach visits (see fig. 12.1). (During each summer an additional fifty teachers also attended the workshop but did not receive the instructional coaching.)

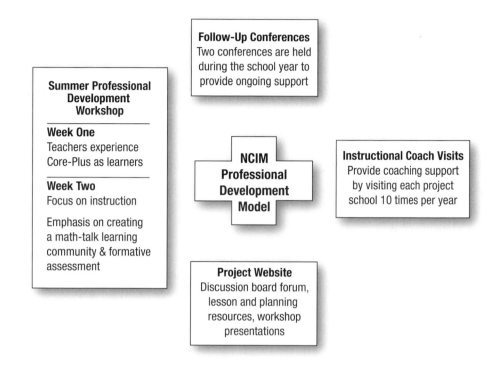

Fig. 12.1. Four components of the NCIM professional development model

Summer Workshops

Teachers implementing reform materials for the first time need firsthand opportunities to experience the curriculum as students working through the investigations (Darling-Hammond and McLaughlin 1995; Schoen and Hirsch 2003; Ziebarth 2001). These experiences enable them to see the significance of the textbook structure and the contributions made by small-group work. We held one-week residential summer institutes in 2008, 2009, and 2010 that focused on using the curricular materials. Based on the evaluation of the 2008 Institute and the reports from instructional coaches and teachers, we added a second week during the 2009 and 2010 summer institutes, one focusing on the pedagogical considerations of teaching reform mathematics. During the 2008–09 academic year, data observation reports showed teachers were still using largely the same traditional instructional practices that they were exposed to as students, confirming reports of similar challenges by others in encouraging teachers to enact reform practices (Ball 1998; Tyack and Cuban 1995).

Web-Based Learning Portal

To support communication and the sharing of resources among the teachers, instructional coaches, project directors, and researchers, a web-based learning portal was created during the second year. Encouraging professional communication can increase the motivation teachers have to continually improve their instructional practices (Garet et al. 2001; Lieberman and McLaughlin 1992) and foster collegiality among colleagues (Loucks-Horsley et al. 2010). The web-based environment had resources for implementing CPMP and a discussion board that created a teacher communication network.

Follow-Up Conferences

The third component of the model included follow-up conferences to the Summer Institute, where teachers came back together for one day in the fall and spring to reinforce and extend the learning that began in the summer. In total, there were four follow-up conferences. Commonly, principals attended these follow-ups in collaboration with teachers at their school; this allowed principals insight on how to offer ongoing support to their teachers in reform mathematics classrooms (Confrey, Maloney, and Krupa 2008; Ziebarth 2001). Topics for the follow-up conferences were selected based on the needs of the teachers, which were self-assessed by teachers and discussed among the project leadership. At the end of the summer institute, teachers evaluated the mathematics content of the curriculum, noted the pedagogical content areas they wanted more exposure to, and created action plans for the upcoming year. Follow-up topics were chosen from common areas that teachers listed on these evaluations. It is important to frequently assess the effectiveness of a professional development model and to make improvements based on the assessments in order to increase teacher effectiveness and student learning (Loucks-Horsley et al. 2010). By assessing the summer institute, project directors created and then improved follow-up conferences based on feedback from the teachers.

The Use of Instructional Coaches

The fourth component of the professional development, and the focus of this chapter, was the monthly site visits from instructional coaches. Two experienced CPMP teachers, Linda and Margaret, conducted the site visits to identify and support teachers' needs. (All names reported here are pseudonyms. Kate was an additional instructional coach for one semester during the spring of 2008.)

This customized mentoring is responsive to individual teacher needs (Darling-Hammond 1997). The NCIM coaching model, designed by the project staff, allowed the coaches to customize professional development during their visits for each NCIM teacher. The NCIM coaching model shared similarities to Loucks-Horsley's et al. (2010) five key elements of coaching: (1) teachers focus on learning or improvement; (2) a climate of trust, collegiality, and continuous growth is cultivated; (3) coaches are well prepared, with in-depth content knowledge; (4) mechanisms for observing practice and providing feedback are critical; and (5) opportunities for interaction are provided. The relationship between the coaches and teachers was formed during the Summer Institute when the instructional coaches acted as facilitators. When the school year started, the coaches determined the needs of each teacher and then addressed them. Coaches were responsible for contacting the teachers to arrange monthly site visits to observe and reflect with the NCIM teachers. Unlike in other models, there was no prescriptive framework for these interactions. The list of activities used to engage teachers will be highlighted and connected to the overall project design below.

A Study of the Design and Use of an Instructional Coach

The participants in this study included twelve NCIM project teachers and the two instructional coaches they worked with throughout the year. Data were collected from classroom observations, semi-structured interviews with both teachers and instructional coaches, and their reports. Instructional coaches completed a report after each of their site visits with teachers to document observations and activities from the visit. The reports provided information on the teachers' use of time and pacing, instructional behaviors, collaborative learning, use of technology, formative assessment, and classroom management. Since the inception of the project in 2008, approximately 170 reports have been completed.

Reports Inform Development of Project

The instructional coaches provided feedback to the directors and workshop developers on what needed to be addressed in other facets of the program. It is difficult to transition from teaching traditional mathematics to CPMP (Arbaugh et al. 2006; Lloyd 2002; Wilson and Lloyd 2000), and the instructional coaches documented the hindrances NCIM project teachers faced when implementing the curriculum. Based on the analysis of the reports submitted by the instructional coaches, the evaluation team identified key themes, and the project directors collaborated to create a summer workshop focusing on these needs. This analysis showed that teachers needed continued support for long- and short-term planning, formative assessment techniques, questioning skills, and content knowledge. Project directors worked together to create a summer workshop focusing on these needs.

The Activities of the Instructional Coach

The instructional coaches engaged in many types of activities with the teachers they supported. As the instructional coaches customized the support they gave each teacher over the course of the project, the list of possible activities has increased to twenty. This activity list then served as the basis for the reports used from the fall of 2008 to the spring of 2010 to produce data on the relative frequency of the different activities. Because the coaching was customized to individual teachers, the percentages are disaggregated by the two coaches to show the variation in the activities they engaged in. The broad categories for reporting, as shown in table 12.1, are (1) curriculum and content assistance; (2) lesson planning, enactment, and reflections, (3) assessment, feedback, and grading; and (4) professional community interactions.

Table 12.1

Percent of total activities of instructional coaches by category

	Total	Linda	Margaret
I. Curriculum and Content Assistance	18.57	16.57	20.29
II. Lesson Planning, Enactment, and Reflection	52.08	44.10	58.94
III. Assessment, Feedback, and Grading	5.71	8.43	3.38
IV. Professional Community Interactions	23.64	30.90	17.39

Then within each category, we report the relative frequency of each of the subgroups within the category (table 12.2).

Table 12.2

Percent of activities performed by instructional coaches within each subcategory

	Total	Linda	Margaret
I. Curriculum and Content Assistance			
Provide additional resources	32.87	47.46	22.62
Discuss or clarify content	30.07	32.20	28.57
Suggest methods for new vocabulary and reading investigations	16.08	5.08	23.81
Offer technical support for technology use	16.08	15.25	16.67
Communicate sequence with teachers and administrators	4.90	0.00	8.33

Table 12.2—*Continued*

	Total	Linda	Margaret
II. Lesson Planning, Enactment, and Reflection			
Assist in daily lesson planning and structure	21.45	20.38	22.13
Offer classroom observation feedback	19.70	26.75	15.16
Foster discourse and questioning	17.46	6.37	24.59
Model teaching	14.21	22.29	9.02
Develop long-term pacing guides	12.72	14.01	11.89
Provide strategies for collaborative learning	9.98	7.64	11.48
Define what a *Core-Plus* classroom looks like	4.49	2.55	5.74
III. Assessment, Feedback, and Grading			
Assist in developing appropriate assessments	27.27	13.33	57.14
Suggest appropriate homework	25.00	30.00	14.29
Develop strategies for providing feedback to students	25.00	26.67	21.43
Assist in developing appropriate rubrics for student work	22.73	30.00	7.14
IV. Professional Community Interactions			
Follow up via email	47.25	46.36	48.61
Promote collaboration	23.63	27.27	18.06
Encourage website use	19.23	18.18	20.83
Communicate needs of teacher	9.89	8.18	12.50

According to interviews with the teachers, some of the most productive activities they engaged in with their instructional coaches were long- and short-term planning, observing the coaches model teaching, getting access to technology and support in their use of it, and receiving feedback following an observation. The case below highlights the specific activities one instructional coach used to influence a teacher's practice.

Customized Professional Development: The Case of Maria

The seven schools receiving instructional coaches have averaged a 27 percent teacher turnover rate per year. High turnover made it difficult to have a strong commitment to implementing the curriculum. The case study presented is evidence that given consistent leadership and a support system, this professional development model can have a profound effect on teachers' instruction.

Maria is one of only two teachers who began in the project during the spring of 2008 and is still teaching integrated mathematics at the same school. She taught traditional mathematics for six years before joining the NCIM project, four of which were in another country where she also taught physics and English. She describes algebra I as her turning point in learning mathematics as a student because it was very hard for her to understand the content. She extensively studied with her father on the weekends to unlock her difficulties with the material. The struggles she faced in mathematics allowed her to relate to students who experience similar difficulties with mathematics and showed her the importance of offering extra support to students after school.

When Maria joined the project, she was the only mathematics teacher at her rural school, and the only teacher in her county teaching CPMP. Her first reaction to the curriculum was:

> This is scary because there is no math here. . . . My fear was, would they (the students) understand these questions, I mean would it make sense to them. That was my initial thought when I started teaching CPMP. I was very fearful with it my first year.

February 2008 marked her first instructional coach visit from Kate. Kate's biggest concern was that Maria's freshman class was two units behind schedule, with a standardized state assessment looming in three months. She also noticed that the students were seated in rows, not working in groups, and there was too much off-task time during transitions. Combining an attempt to use a reform curriculum with traditional pedagogical practices seemed to be exacerbating Maria's difficulties with pacing.

After the visit, Kate developed a coaching plan for Maria. Along with two other teachers experienced with CPMP, Kate created a long-term planning guide aligned with Maria's district calendar. She collected resources that Maria could use in her classroom: an overhead, a timer to help with transitions, and supplies to use with the investigations in the curriculum. In April, Kate brought Maria to an urban school district where teachers had been teaching CPMP for five years. Together they watched four classes. This was the first time Maria was able to experience a reform mathematics classroom in situ. Kate made sure each of the classes provided a different perspective, and she reported on the benefits of this experience:

- "Maria is seeing the true picture today. The teacher is talking about geometric figures with several set side lengths or angle measures. She has toys that help her students visualize the problems. She has a car jack (fig. 12.2) that is like the photo in the textbook. Maria sees how much the students respond to this kind of stimulus. The teacher also combines some large-group discussion with small-group investigations. Maria sees the vision."

- "We visit a class where the kids are noisy and not always on task. This looks familiar to me, and Maria realizes this is something like her class. We can see the strategies the teacher uses to pull kids back into the task. We can also see what happens if they get further and further off task. This was a good choice."

- "Next we go to the fourth-year course. Maria can see her students three years from now. She is totally blown away about the math they are doing—exponential functions, and they are doing investigations! Look at how focused they are; they go right to work and stay on task. The vision is there."

- "The teachers all use technology. Two of them have Smartboards. Maria looks at me with her mouth wide open. She has never seen this. One of the teachers uses some graphs from the teacher notes (on CD) to help her students see a graph of the physical phenomenon that the teacher is showing the students by physically moving the parts of the triangle. So many good ideas."

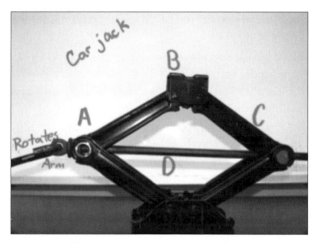

Fig. 12.2. Car jack manipulative used for instruction

As the only mathematics teacher at her school, Maria had never observed a CPMP classroom, and this visit allowed her to envision the reform mathematics environment Kate had been endorsing. The experience also highlighted the strengths and weaknesses of the observed teachers' instruction, which Maria used to reflect on her own practices. After this visit Maria spoke with her principal about getting an LCD projector in her classroom, an initial step in transforming her teaching practices.

At the principal's request, Maria's new instructional coach, Linda, returned the next year (fall 2008) for a two-day visit. Linda was a new coach and would be a facilitator at the 2009 summer institute. She believed in the curriculum so intensely that she had successfully implemented CPMP into her traditional classes. Her favorite activities to engage in with her project teachers were model teaching and helping small groups, as she enjoyed interacting with the students. Table 12.3 shows the percentage of activities Linda used with Maria within each broad category during the 2008–09 and 2009–10 school years.

Table 12.3

Percent of total activities Linda used with Maria

	Total
I. Curriculum and Content Assistance	12.90
II. Lesson Planning, Enactment, and Reflection	51.61
III. Assessment, Feedback, and Grading	4.83
IV. Professional Community Interactions	30.65

Fig. 12.1. Four components of the NCIM professional development model

When Linda came for her first observation, Maria's class was working in groups, and she had an LCD projector displaying CPMP-Tools, software designed for the curriculum. Maria had put a lot of time into organizing her classroom and establishing group roles and norms to be used with her students. Maria began to take ownership of the teacher-coach relationship, and she requested that Linda come to visit during a teacher workday so they could have a full day for collaborating.

At the beginning of the spring 2009 semester, Maria and Linda began by creating a pacing guide for CPMP Course 2, and they posted it on the project's web-based environment for other teachers to use as a resource. During this meeting Linda also gave suggestions related to some concerns she had reported during her previous observations. While Maria was beginning to incorporate groups, she still had her students seated in rows at the beginning of the class and the end of each period. When it was time for group work, the students traveled around the classroom to form their groups; this was a very slow process and some of the group dynamics were ineffective. Linda made three suggestions to help with the group: Use a timer between transitions (http://www.shodor.org/interactivate/activities/Stopwatch/); seat students in the same group closer to each other; and use a matrix model for heterogeneously grouping students by ability level (see fig. 12.3). Using this grouping model, the teacher creates a matrix with the columns equal to the number of desired groups and writes the names of students of high ability in the first row and low ability in the last row. The columns determine the groups, creating groups of students with mixed high, medium, and low abilities. Based on these suggestions, Maria made attempts to improve her instruction so that students would not be off-task. The type of encouragement and scaffolds Linda provided are given in Linda's September 2009 reflection to Maria:

> You continue to do a great job using the timer to keep you and the class on track. Your use of the data projector adds a lot to your instruction. . . . You are patient with them and try to nudge them toward the right answer without just giving up and answering the question for them. I really like that you have the students write down the essential question from the investigation with the idea of coming back to answer it at the end of the investigation. I thought you did a great job with the launch to review solving an equation and providing reasons for each step (proof). You are always good about asking the students if they have any questions before you move on to the next activity. I am so glad that you are getting your students to sit near their groups. It is amazing how much time you can save. I still think they could do it a little faster but you can work on it. You have done such a good job training them. They immediately got into their groups and began working. I liked that you got students to put work up on the board and to explain it to their classmates.

Ability	Group 1	Group 2	Group 3	Group 4	Group 5	Group 6
Highest	*H*	*H*	*H*	*H*	*H*	*H*
Medium	*M*	*M*	*M*	*M*	*M*	*M*
Medium	*M*	*M*	*M*	*M*	*M*	*M*
Low	*L*	*L*	*L*	*L*	*L*	*L*

Fig. 12.3. Technique for heterogeneously grouping students

As a participant of the NCIM project receiving an instructional coach for two and a half years, Maria made major changes to her teaching practice. She transitioned from direct instruction to an environment where students generated and shared their ideas. Maria was able to incorporate collaborative grouping more effectively, ensuring that students were engaging in rich discussions about mathematics. The increased use of technology allowed Maria to use time more effectively during transition periods, introduce students to real-world problems, and teach them twenty-first-century skills using computer-based software and graphing calculator technologies.

These changes resulted from Maria's willingness to accept constructive criticism from her instructional coaches, geared toward improving her teaching practices. She now views her role in the classroom as a facilitator and understands how important it is to get students to collaborate with one another on mathematical content. Maria points out that to get students to rely on one another she "goes around asking them questions, and every time they ask me a question, I ask them another question." She attributes the changes she has made to the support of the instructional coaches, her attendance at the workshops, and the resources on the project web page. In addition, Maria has a very supportive principal who attended facets of the professional development along with her, trusted the instructional coaches to work with her without interference, and provided her with professional leave for the conferences and site visit with Kate.

In the case of Maria, the benefits of the instructional coach's customized support resulted in additional positive effects on other teachers. As a teacher, Maria became an unofficial instructional coach herself as new teachers entered the program. Two new mathematics teachers, Heather and Nora, were hired at Maria's school in August of 2009, both unfamiliar with reform mathematics. All three teachers receive support from Linda, but Maria also modeled with the new teachers those coaching strategies her instructional coaches had used to help her become successful. During the first two months of school, Nora observed Maria's first two periods and then Maria went to observe Nora's fourth period. This gave Nora the opportunity to observe a demonstration of effective teaching practices, reflect on them during lunch, practice them during her lesson, and receive feedback from Maria. The activities that Maria did with Linda helped her transition from an apprentice to a mentor for her colleagues. This collaboration helped Heather and Nora strengthen their instructional practices and helped to scale the project's mentoring model. Of the twelve teachers receiving an instructional coach, Maria was one of two teachers to model coaching techniques with their new colleagues. This modeling has not happened with the other teachers because two teachers were the only mathematics teachers at their school and seven teachers were teaching the material for the first time.

Conclusions

Considering her initial trepidation when it was mandated she teach integrated mathematics, Maria's case is encouraging. When provided with support, she was able to transfer the knowledge and skill she learned from her instructional coach and during workshops into practice. Other teachers also had a noticeable change in their practice, especially those in their first three years of teaching. As a whole, the twelve teachers improved most notably in the effective use of class time, long-term pacing, formative assessment techniques, and content delivery (Krupa and Confrey 2010). Maria's case was presented because she had the longest relationship with her instructional coaches and the most time to improve her instruction. Though not all participating teachers experienced this level of success, Maria's case represents the benefit of working one-on-one with an instructional coach. It is also important to realize the changes in Maria's instructional practices were gradual, and the treatment with the instructional coach took time.

Principles and Standards for School Mathematics (NCTM 2000) provides an image of thriving mathematics classrooms that present mathematics problems through a variety of topics, solve them with applications rich in technology, and show their solutions in multiple perspectives. Regardless of the curriculum used to facilitate mathematical understanding, teachers need to be supported to create this type of environment. With CPMP, teachers' beliefs change as they transition from teaching traditional mathematics, and this transition takes time and requires support (Arbaugh et al. 2006; Lloyd 2002; Wilson and Lloyd 2000). One model that has helped teachers like Maria with this transition is providing them with an instructional coach. For the NCIM project, the model is sustainable because teachers are supported throughout the school year by instructional coaches, which, at times, led to teachers becoming more supportive of one another. This model needs to be replicated across different sites to validate this approach. Over time, by incorporating reports from coaches and suggestions from teachers, project directors created a professional development in the context of teachers' classrooms. This type of coaching is aligned with Guskey's (1999) argument that "professional development is an ongoing activity [which should be] woven into the fabric of every educator's professional life" (p. 38).

REFERENCES

Arbaugh, Fran, John Lannin, Dustin L. Jones, and Meredith Park-Rogers. "Examining Instructional Practices in Core-Plus Lessons: Implications for Professional Development." *Journal of Mathematics Teacher Education* 9, no. 6 (2006): 517–50.

Ball, Deborah Loewenberg. "Teacher Learning and the Mathematics Reforms: What We Think We Know and What We Need to Learn." *Phi Delta Kappan* 77, no. 7 (1996): 500–508.

Ball, Deborah Loewenberg. "Unlearning to Teach Mathematics." *For the Learning of Mathematics* 8, no. 1 (1998): 40–48.

Ball, Deborah Loewenberg, and David K. Cohen. "Developing Practice, Developing Practitioners: Toward a Practice-Based Theory of Professional Education." In *Teaching as the Learning Profession: Handbook of Policy and Practice*, edited by Linda Darling-Hammond and Gary Sykes, pp. 3–22. San Francisco: Jossey-Bass, 1999.

Confrey, Jere, Alan Maloney, and Erin E. Krupa. *N.C. Integrated Mathematics NC-STEM MSP: Year-End Evaluation Report.* Raleigh, N.C.: North Carolina State University, 2008.

Darling-Hammond, Linda. *The Right to Learn: A Blueprint for Creating Schools That Work.* San Francisco: Jossey-Bass, 1997.

Darling-Hammond, Linda, and Milbrey W. McLaughlin. "Policies That Support Professional Development in an Era of Reform." *Phi Delta Kappan* 76 (1995): 597–604.

Darling-Hammond, Linda, Ruth Chung Wei, Alethea Andree, Nikole Richardson, and Stelios Orphanos. *Professional Learning in the Learning Profession: A Status Report on Teacher Development in the United States and Abroad.* School Redesign Network at Stanford University and the National Staff Development Council, 2009.

Garet, Michael S., Andrew C. Porter, Laura Desimone, Beatrice Birman, and Kwang Suk Yoon. "What Makes Professional Development Effective? Results from a National Sample of Teachers." *American Educational Research Journal* 38, no. 4 (2001): 915–45.

Guskey, Thomas R. *Evaluating Professional Development.* Thousand Oaks, Calif: Corwin Press, 1999.

Guskey, Thomas R. "Professional Development and Teacher Change." *Teachers and Teaching Theory and Practice* 8, no. 3 (2002): 381–91.

Joyce, Bruce, and Beverley Showers. *Student Achievement through Staff Development.* 3rd ed. Alexandria, Va.: Association for Supervision and Curriculum Development, 2002.

Krupa, Erin E., and Jere Confrey. "Teacher Change Facilitated by Instructional Coaches: A Customized Approach to Professional Development." In *Proceedings of the 32nd Annual Meeting of PME-NA.* Columbus, Ohio, 2010.

Lieberman, Ann, and Milbrey W. McLaughlin. "Networks for Educational Change: Powerful and Problematic." *Phi Delta Kappan* 73 (1992): 673–77.

Lloyd, Gwendolyn M. "Mathematics Teachers' Beliefs and Experiences with Innovative Curriculum Materials." In *Beliefs: A Hidden Variable in Mathematics Education?,* edited by Gilah C. Leder, Erkki Pehkonen, and Günter Törner, pp. 149–59. Dordrecht, the Netherlands: Kluwer, 2002.

Loucks-Horsley, Susan, Katherine E. Stiles, Susan Mundry, Nancy Love, and Peter W. Hewson. *Designing Professional Development for Teachers of Science and Mathematics.* 3rd ed. Thousand Oaks, Calif.: Corwin Press, 2010.

National Council of Teachers of Mathematics (NCTM). *Principles and Standards for School Mathematics.* Reston, Va.: NCTM, 2000.

Putnam, Ralph T., and Hilda Borko. "Teacher Learning: Implications of New Views of Cognition." In *International Handbook of Teachers and Teaching, Vol. 2,* edited by Bruce J. Biddle, Thomas L. Good, and Ivor F. Goodson, pp. 1223–96. Dordrecht, the Netherlands: Kluwer, 1997.

Schoen, Harold L., and Christian R. Hirsch. "The Core-Plus Mathematics Project: Perspectives and Student Achievement." In *Standards-Based School Mathematics Curricula: What Are They? What Do Students Learn?,* edited by Sharon L. Senk and Denisse R. Thompson, pp. 311–44. Hillsdale, N.J.: Lawrence Erlbaum Associates, Inc., 2003.

Stein, Mary Kay, Margaret Schwan Smith, and Edward A. Silver. "The Development of Professional Developers: Learning to Assist Teachers in New Settings in New Ways." *Harvard Educational Review* 69, no. 3 (1999): 237–70.

Tyack, David, and Larry Cuban. *Tinkering toward Utopia: A Century of Public School Reform.* Cambridge, Mass.: Harvard University Press, 1995.

Wilson, Melvin S., and Gwendolyn M. Lloyd. "Sharing Mathematical Authority with Students: The Challenge for High School Teachers." *Journal of Curriculum and Supervision* 15, no. 2 (2000): 146–69.

Wilson, Suzanne M., and Jennifer Berne. "Teacher Learning and the Acquisition of Professional Knowledge: An Examination of Research on Contemporary Professional Development." *Review of Research in Education* 24, no. 1 (1999): 173.

Ziebarth, Steven W. "Prime-Team: The University of Iowa Local Systemic Change Project." *Mathematics Teacher* 94, no. 4 (2001): 336.

CHAPTER 13

The Role of Communities of Practice in Developing Teacher Leadership

Antonia Cameron
Frances Blanchette
Glenda Francis
Christina Fuentes
Mayra Rivera-Deliz

THIS CHAPTER EXPLORES A MODEL of professional development in which teacher leadership and improvements in classroom practice are directly linked. We will begin by tracing the origins of our community in the individual words of the chapter's five co-authors. After that, we will follow the journey of one teacher in our learning community as she improved her practice. In the concluding sections, we reflect on our learning as members of this community.

Origins of the Learning Community: Questions Posed

The following five sections introduce the voices of key members of our learning community: a professional staff developer, two teacher leaders, one principal, and one math coach. (We are defining a teacher leader as someone who serves as a model or support to other teachers in her school or district.) We do this to highlight how the questions we each initially sought to answer were the starting point for our collaboration to develop teacher leadership in mathematics.

Toni, a Professional Staff Developer

My question: *How might a classroom-based learning community be used to simultaneously improve teacher practice* and *develop teacher leaders?* This question emerged from two insights

I had as a staff developer for the Mathematics in the City project (MitC). (This project [www.mitccny.org] is a national center of professional development for K–8 mathematics education located at the City College of New York.) First, I noticed that many workshops for teachers—even those facilitated by highly skilled and respected mathematics educators—only minimally affected teacher practice. As I visited classrooms, I saw firsthand that what teachers implemented from workshops was often a reinterpretation or a misinterpretation of what they had learned, a watering down of powerful mathematical ideas and pedagogies (Lieberman and Miller 2008).

Second, from consulting in many different classrooms and schools, I knew that working classroom-by-classroom and school-by-school would never create deep or lasting change. What I was doing was *systematic* (that is, I was using the same co-teaching model with the individual teachers I was coaching), but it was not *systemic*. To create systemic change, I needed to find a different model for professional development, one that had the power of on-site co-teaching and collaboration but was also able to influence a larger number of educators. Additionally, for the change to be systemic, the professional development model needed to be self-generating—a model in which, at some point, I was no longer needed. I therefore focused my work on creating a network of teacher leaders.

The general need for a new professional development model focused on developing teacher leaders has led to the creation of communities of practice (Wenger 1999). These have existed in a variety of forms: school-based, district-based, and city-wide. The Collaborative Communities of Practice (CCP) and the MitC Learning Communities were two examples of such communities. (The CCP was a project developed by the NYC Department of Education; MitC ran the mathematics component of these communities.) Both projects focused on (1) developing leaders across school roles (coach, teacher, and administrator); (2) creating lab-site schools and classrooms (these were the collaboration sites that hosted the learning communities); and (3) using discourse as the primary tool to engage and challenge participant thinking (Herbel-Eisenmann and Cirillo 2009). Each collaboration also used a co-teaching model of professional development. This meant that I (Toni) coached the classroom teacher (or coach) in the actual moment of teaching (or coaching) (West and Staub 2003; West and Cameron, in press).

In these learning communities, educators opened their practice to visiting teachers, coaches, and administrators. Hosting teachers would talk publicly about their teaching and invite visiting educators into their classrooms to observe them co-teaching math lessons (Stein, Silver, and Smith 1998). Visiting educators participated in planning and debriefing the lesson, and they also took notes during classroom visits on the teacher's facilitation of the lesson and the role of student communication in learning. These notes were used to deconstruct the lesson in the debriefing session and to create a coherent vision of mathematics teaching and learning.

Later in this chapter, we will see an example of how a learning community supports and develops teacher leadership by using one school, PS 24, to highlight one teacher leader's journey. (PS 24, Brooklyn, is a NYC public elementary school. Of its 770 students,

95 percent are from low-income families, and approximately 50 percent of the ELLs live in the mostly Latino, immigrant community of Sunset Park.) First, though, let's hear from the other key members of our collaborative community.

Glenda, Teacher Leader I, PS 24

My question—*Why learning communities?*—was answered when I discovered how radically different they are from other professional development experiences. I had never participated in a structured, collaborative forum that *connected adult learning to student learning*—a place where, for example, a mathematics problem explored by adults in the morning was later presented to students in the classroom. How surprising it was that our struggles were their struggles, that our ah-has were their ah-has! How powerful it was that the facilitation of our thinking mirrored theirs, too. We were challenged to create, justify, revise, and reflect on our learning, and so were they (Lampert and Ball 1999).

What became clear to me was that this community gave adult learners opportunities to construct mathematical ideas related to the content they were expected to teach and to bump against their previous understandings. It was transformative because teachers were challenged to think in new ways about their practice and to share their ideas publicly.

Frances, Teacher Leader II, PS 24

My question—*How can I be a more effective teacher?*—grew out of my struggles as a beginning teacher. Glenda had just come to PS 24, and whenever I passed her classroom I felt an emotional and intellectual life resonating within her classroom. Her students exuded such confidence. I approached Glenda with questions about her practice, and, to my delight, she welcomed my inquiries. Her openness led to ongoing conversations about teaching and learning.

Because Glenda was generous with her time and knowledge, I learned a great deal about classroom setup, how to empower students, and the critical role discourse plays in developing student thinking (Alexander 2004). Working with Glenda made me keenly aware of the power of collaboration. Listening to her voice helped me find my own, and her open door became my open door. Together we invited other teachers into our conversations as part of the MitC learning communities.

Christina, Principal, PS 24

My questions were intertwined: *How could I support the great teacher synergy I saw developing between Glenda and Frances?* and *How I could use this collaboration to support the learning of other teachers at PS 24?* I put their classrooms side by side, so that cross-fertilization could easily occur. I offered them many opportunities to participate in mathematics professional development together. Their collaboration made them ideal candidates to host MitC learning communities. This collaboration led to other teacher collaborations, which flourished as more teachers became immersed in this work.

Mayra, Math Coach, PS 24

My question was: *How do we support a community of adult learners at PS 24?* Christina and I had participated in MitC learning communities and we knew how powerful these could be in affecting change. When the opportunity arose to host one, we seized it. But none of this was possible without creative uses of resources (time, money, and scheduling) and the development of systems (i.e., common planning time) to foster teacher collaboration. One way Christina supported teacher learning was to build in time after each community for participants to share their learning. This meant that Christina provided classroom coverage to grade-level teams for planning and inter-classroom visitations.

The Collaboration of Toni, Glenda, Frances, Christina, and Mayra

Our collaboration melded our individual needs, passions, and questions. For our community to thrive, we needed to believe that everyone had something important to contribute to our collective work (Louis and Kruse 1995). The community could not have functioned without the careful attention to organizational details provided by Christina and Mayra; nor could it have survived without teacher leadership—Glenda and Frances's willingness to publicly reflect on their teaching in conversations facilitated by Toni. Our goals were twofold: (1) to develop teacher leaders whose practice would be the focus in discussions about the teaching and learning of mathematics; and (2) to create networks of teachers to continue the conversation long after the MitC learning community ended. However, the heart of our collaboration was the development of teacher leadership.

Why teacher leadership? For us, creating changes in mathematics teaching and learning is directly linked to improving classroom practice. One of the most effective ways to support this needed change is to have teachers who are able to influence their peers. To do this, however, teacher leaders need a complex set of skills. These include being able to collaborate and open their practice to others, able to facilitate adult learning, and able to challenge opinions and be comfortable with the disequilibrium that ensues. Teacher leaders also need content and pedagogical content knowledge deep enough to be able to unpack and teach a mathematics lesson (Ball, Thames, and Phelps 2008).

In the remainder of this chapter, we will explore how our collaboration supported the development of teacher leadership within and across schools (York-Barr and Duke 2004). Specifically, we will use transcripts of discourse in our community to trace the professional development of Frances, a teacher leader at PS 24. We will also share the reflections of Melissa, a participating teacher from another school, whose growth can be directly linked back to her experiences in a learning community hosted by Frances.

Discourse in the Learning Community

In this section, we highlight transcripts from two different learning communities as a means for illustrating (1) how Toni's facilitation of discourse in our community of adult learners supported and challenged teachers' thinking about student work and proof (transcript 1); (2) how Glenda, an experienced teacher leader, and Toni refocus and elevate the whole-group discussion while honing Frances' pedagogy (transcripts 1–3); (3) how Frances' public reflections generate changes in her practice (transcripts 2 and 3); and (4) how the goals Frances sets for improving her practice ultimately impact student learning in her classroom (transcripts 5–7).

Cultivating a Teacher Leader

The first three transcripts are taken from a fifth-grade learning community, co-hosted by Glenda and Frances and facilitated by Toni. We begin by illustrating Frances's journey as a teacher leader. In transcript 1, the team is analyzing student work in preparation for a math congress, which will be co-taught by Toni and Frances. (A math congress differs greatly from the norms for sharing children's solutions seen in most classrooms. In a math congress, specific children's ideas are presented before the mathematical community, and the congress is carefully designed so that student ideas, modeling, and/or questions become the tools for extending children's thinking.) The discussion here is about the "The Box Factory," a unit in *Context for Learning Mathematics* (Jensen and Fosnot 2007). The conversation is focused on children's strategies and whether or not their posters show that they have found all the possible rectangular box arrangements for twenty-four chocolates (see fig. 13.1).

All the ways to make boxes of 24	**How we know we have them all.**
1. $(24 \times 1) \times 1$ 2. $(2 \times 12) \times 1$ 3. $(6 \times 4) \times 1$ 4. $(3 \times 4) \times 2$ 5. $(6 \times 2) \times 2$ 6. $(12 \times 1) \times 2$ 7. $(1 \times 8) \times 3$ 8. $(4 \times 2) \times 3$ 9. $(3 \times 2) \times 4$ 10. $(1 \times 6) \times 4$ 11. $(2 \times 3) \times 4$ 12. $(1 \times 2) \times 12$ 13. $(3 \times 1) \times 8$ 14. $(2 \times 2) \times 6$ 15. $(4 \times 3) \times 2$	How do we know we have them all is that we thought about the factors of 24, 6, 2, 8, 4, 3, 1, 12. We got all of the factors of 24, 6, 2, 8, 4, 3, 1, 12, and then we started mixing them so they could give us boxes of 24. Then when we were done we maked [sic] sure we had all of them.

Fig. 13.1. Transcribed text from student poster under discussion by teachers

TRANSCRIPT 1—

Teacher: This work is consistent; they used the same factors, 3, 1, and 8. That's systematic.

Frances: And they created cube models of the boxes, so if we ask them to speak, they will have their models to show.

Glenda: The question *Have you worked systematically?* is hard. Because doubling and halving is systematic. If I'm flipping, that's systematic. So that a consistent use of the strategy is one way to be systematic and then how do you organize it to show . . .

Toni: So you can have a system, but your system is not necessarily a proof . . .

Glenda: Your system has to be systematized!

Teacher: So how would kids know they have all the possible ways?

Frances: Could *we* prove we have all the possible ways?

Toni: In my experience, no, because how much experience do we have with these kinds of questions—what does it mean *to prove your thinking?*

In this brief segment, Glenda names the strategy—doubling and halving—that students are using to solve the task. This begins a discussion about the difference between a systematic strategy and a proof (Shifter 2009). When Toni emphasizes that a "system is not necessarily a proof," she wants teachers to think about their *own* experiences with proof. In response, a teacher poses a question about student solutions and proof that actually hints at a problem confronting the group: *If we don't know how to prove we have all the possible ways, how could we recognize a proof in student work?* When Frances asks, "Could *we* prove we have all the possible ways?" we see her emergence as a leader. Frances understands and uses the subtext of a teacher comment to challenge the learning of the group. She recognizes that the real issue is not *student proof*, but *teacher understanding*. Her reframing of the teacher's question shifts the adult learning by refocusing the lens they are using for analyzing student work. This challenge has the potential to create disequilibrium in the group, but as a leader, Frances takes that risk.

The conversation in transcript 2 takes place after teachers have visited Frances's classroom and viewed a math congress. This particular congress was designed to juxtapose students' solutions (shown on posters) with their justifications for saying they had found all the possible box arrangements for twenty-four chocolates.

This kind of sharing is enormously difficult for teachers to facilitate because of the challenges associated with helping structure discourse around children's ideas and confusions (Brent and Simmt 2003). Often, as teachers deal with student confusions, they lose the mathematical focus (Fosnot and Dolk 2001).

TRANSCRIPT 2—

Toni:	How do you feel after your congress? [*The debriefing session usually begins with the teacher and facilitator starting the discussion. This gives the teacher a chance to reflect publicly before others share their noticings.*]
Frances:	Confused!
Toni:	Why confused?
Frances:	In the end, I'm not really sure what the kids got out of it.
Toni:	Perhaps the conversation at the end of the congress needed to be more precise.
Glenda:	Exactly. We need to say, "We've shared so many ideas that I'm feeling confused. Could we name something that we got from this congress and what we're still confused about?"
Frances:	And see what kids say?
Toni:	Or send them off to write in their journals. You can say something like, "We had a lot of confusion; some really important insights—what are *you* taking away from this math congress?"
Glenda:	Dina was a good model. She said, "Now I'm changing my ideas." You could say something like, "Wow! This is what our congress is designed to do. We have some ideas. We're reporting to a community of mathematicians and we're forced to defend or revise our thinking."

In this segment, Toni and Glenda help Frances think about why confusion is important for learning. They outline three things crucial to learners: (1) emotions occur in learning and are nameable (such as *I'm confused*), (2) self-reflection develops thinking (*What do I understand from this discussion?*), and (3) defending and revising one's thinking (*I'm changing my thinking now*) is part of the process. They also name specific ways Frances can help students reflect on their learning (e.g., journal writing).

The post-lesson conversation continues as Toni and Glenda name specific pedagogical tools that teachers can use to slow the conversation and help children process the ideas being presented (Chapin and O'Connor 2007).

TRANSCRIPT 3—

Frances: I struggle to keep my focus on the big ideas and integrate what kids say into that.

Toni: The share and pulling children's ideas together is probably the most difficult thing. It's so messy. As I was facilitating, I was listening for certain things and I went with them. What moves facilitated the conversation? It might help to think about when I wrote on the board, what I wrote, and why I chose to write at that moment. When did I slow things down with pair talk?

Glenda: When did kids show the models? That was a turning point. We're not talking in the abstract; here's what we're talking about. Now we can manipulate this thing and use it as evidence, as Roberto said.

Toni: These are important pedagogical tools, but then you have to internalize them into your practice.

Frances: It's not that I don't do these things; I guess it's just recognizing when to do it. That takes practice. It's having the ideas in your head, being ready, and having the presence of mind to focus in on what kids are saying and interpret it. And I'm just not there yet … This is something I really want to work on and need help with.

In this transcript, Frances names her struggle: how to keep the discussion focused on mathematical ideas and simultaneously entertain student confusion. Toni and Glenda name specific pedagogical tools (wait time, pair talk, and the use of board space and mathematical models to illuminate key ideas) that will help Frances keep a mathematical focus and deal with students' needs for processing ideas. Two of these tools—wait time and pair talk—create thinking space for students by slowing down the conversation. They also give children a chance to reflect on and communicate their understanding with their peers (Choppin 2007).

The Emergence of a Teacher Leader

That Frances integrated the pedagogical moves suggested by Glenda and Toni becomes evident in the next four transcripts. These narratives were collected in her classroom the following year. Here Frances is facilitating a math congress before a visiting fourth-grade learning community. (It is important to note that Toni is not co-teaching with Frances in this math congress; this is a major shift in her development as a teacher leader.)

TRANSCRIPT 4—

Frances:	Who wants to share what they remember from yesterday's congress? [*Uses wait time*] Ana?
Ana:	One idea we talked about is that all the even multiples of 3 that are factors of 24 are factors of 48.
Frances:	Who can remember that? [*Few hands are raised*] Let me write this on the board to help you remember. This was a conjecture, right? So the conjecture was that [*writes*] "all the even multiples of 3 …"
Ana:	… that are factors of 24 are factors of 48.
Frances:	Thumb on your chest when you remember some of the things we talked about yesterday. [*Scans the room and waits for children's thumbs*] Okay, it seems like at least one person in each partnership is starting to have some memories. Talk to your partner about what you remember from yesterday's congress.

In this transcript, Frances begins her congress by asking students to reconstruct yesterday's learning (a conjecture created by another student). She uses two tools (wait time and writing) to support children's learning. Careful scaffolding like this sets the stage for a new idea to enter the community and for children to grapple with it.

The conversation in the following segment occurred after two children, Mela and Jose, presented their strategy to the community (see fig. 13.2). Specifically, the students were discussing how the factors of 24 are also factors of 48; once you know the factors of 24, you double them to know the factors of 48 (e.g., 1×24 becomes $2 \times 1 \times 24$ or 2×24 or 1×48).

TRANSCRIPT 5—

Frances:	Let's take a minute to process what Mela just said. Raise your hand if you understand her [*Some do*]. Raise your hand if you're confused [*Others do*]. Could someone paraphrase this idea? Yesenia?
Yesenia:	What Mela is saying is that if she knows that 24 is the half of 48, then all the factors of 24 will fit equally on the other half of 48.
Frances:	[*To Mela*] Is that what you're saying? [*To the class*] Have we seen this idea before, or is this a new idea?
Children:	New idea!

TRANSCRIPT 5—Continued

Frances: This is a new idea for us. So let's take a minute to process this with our partners. Turn to the person next to you, and try to make some sense of what Mela and Jose are saying.

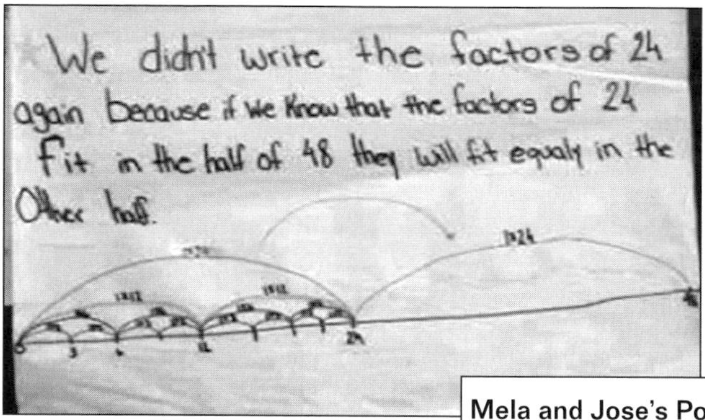

We didn't write the factors of 24 again because if we know that the factors of 24 fit in the half of 48 they will fit equaly in the other half.

Mela and Jose's Poster

Fig. 13.2. Student work of poster being used by Frances in her math congress

In this transcript, Frances anticipates and acknowledges student confusion and uses wait time, pair talk, and paraphrasing—key discourse tools—to support learning. Frances also names Mela and Jose's strategy as a new idea and highlights a critical learning behavior: When new ideas are encountered, learners need time to process them (Cobb, Wood, and Yackel 1993).

After pair talk, Frances brings other children into the conversation that follows. Including more children in classroom discourse is a way for Frances to check student understanding and deal with confusions still in the community.

TRANSCRIPT 6—

Frances: Who would like to give back to the community about this idea? David?

David: What I understand is that all the factors of 48, if you cut them in half, they could fit in the other half?

Mela: [Shakes her head no.]

TRANSCRIPT 6—Continued

Frances: [*To Mela*] That's not what you're saying? Okay, so speak back to David about this.

Mela: We said that the factors of 24 fit in this half since you know that this 24 is the half of 48.

Frances: David, does that make sense to you?

David: [*Shakes his head no.*]

Frances: That doesn't make sense to you. Good for you for being aware of that! Can anyone speak back to this idea that David and some of the other folks are still confused about?

Frances focuses on children's understanding; this is evident in her facilitation of their talk. She does not correct, paraphrase, or lead the conversation. The flow of the discourse is generated by student ideas and confusions. Because she celebrates self-awareness ("Good for you for being aware of that!") and risk-taking, her children rise to her expectations to be public learners (Costa and Kallick 2000).

Following this, much time is spent discussing Mela and Jose's strategy. At the end of the congress, a different student offers a conjecture.

TRANSCRIPT 7—

Frances: Javier, do you think you could explain Ana's idea?

Javier: What Ana is saying is that every factor that's in 12—like 3 is in 12— you just double it and everything will be the same.

Juan: It's like a copy.

Frances: It's like a copy, huh? Ana, are you saying that if we look inside the 24 that we can see the same idea we're talking about in 48? Because 12 is half of 24, I can look at this same idea and think about the factors of 12 fitting into 24. Oh, my goodness! Would this work with other numbers? [*Long pause*] Okay, this is a good place to stop. Today some new ideas came into our community. I want to give you time to synthesize them, reflect on them, and write about them in your math journals. What are some of the new ideas and understandings that came up for you and what are you still not sure about?

Frances recognizes that Ana's conjecture—her leap in thinking is too broad for many of her students to follow (Reid and Zack 2009). She uses their language ("It's like a copy") and chooses an opportune moment to end the congress by having children reflect on their learning by writing in their math journals. This facilitation move was suggested in the third narrative. There, Toni suggested that student writing could be used as a tool to simultaneously help students reflect on learning and give the teacher insights into student understanding.

In these final four short transcripts, we can trace Frances's development. She has developed specific content and pedagogical content knowledge that she is using in her own practice. She has developed ways to notice student behaviors and can now use these noticings to support learning (Sherin, Jacobs, and Philipp 2011). Her powerful facilitation of mathematical ideas, her comfort with the messiness of children's learning, and her use of important pedagogical tools to support discourse can be directly linked to the work of the previous year's learning community and to the goals Frances set for herself as a teacher. From this collaboration, a teacher leader has emerged.

Reflections on Teaching and Leading: Frances, Glenda, Mayra, and Christina

We can unequivocally say this learning community had significant and lasting impact on our teaching and leading, and on the PS 24 school community. We think this happened for several reasons. First, we rooted our learning in conversations about our practice. Second, by publicly sharing our practice and acknowledging our weaknesses, we were able to grow (Darling-Hammond and Richardson 1999). At our school, teachers who participated in our learning community became more willing to share their practice (i.e., become leaders). As teachers opened their doors and invited their colleagues in to watch them teach, they brought new teachers into the conversation (Stein, Silver, and Smith 1998). These collaborations and conversations have also helped us establish clear goals and standards for children's mathematical learning in our school. Third, as a community we got used to, and even embraced, having our ideas and beliefs challenged. Christina writes:

> As Frances gained experience as a teacher leader, her meticulous, supportive classroom environment and her disciplined teacher practice were an example for the faculty. However, her influence on others was significant in more unexpected ways. For instance, once she became involved in a heated debate with a math cluster teacher over the value of teaching the traditional algorithm. Many faculty members became aware of this debate and a healthy whole-school conversation ensued. Certainly, Frances's willingness to speak up, and her fearlessness in rocking the boat, had a lasting positive effect on the culture of our school.

Reflections on Teaching and Leading: Toni

For me, teacher transformations occur because connections are being made among individuals in our learning community as they share ideas, ask difficult questions, and challenge each others' beliefs and practices. That these connections are deep and live on beyond the life of the community is demonstrated in the reflections of Melissa, a participating teacher in the fourth-grade learning community:

> I am currently serving as a teacher leader co-facilitating a MitC learning community at PS 230. Several years ago, I attended a learning community at PS 24 co-facilitated by Frances and Toni. This experience changed my thinking about what it means to do and teach math. I had never heard a teacher talk so openly and passionately about what she did in her classroom, or thought it possible that a single math lesson could be planned in such depth and with such a discerning eye for student learning. I had never been in a classroom where I saw students empowered by their own ideas speaking clearly about those ideas and defending them. I remember thinking, "This is what I want my classroom to look like; this is how I want my students to be as learners." One of my biggest insights was that Frances herself was the model for student learning! The expectations she had for her students she had for herself. Just as Frances' generosity changed my teaching, that's how I want to collaborate with others and pass on to them what was given to me.

This quote reflects the generative power of teacher learning in a collaborative community. One teacher's passion for learning generates another teacher's enthusiasm. A teacher's excitement and willingness to publicly learn has the potential to both transform her practice and impact student development. A teacher's willingness to take risks—to open up her practice for observation and critique—reflects a major change in mindset, a movement from focusing solely on her own teaching to thinking about how to influence and support the teaching of others (Little 1999). This shift in perspective exemplifies the journey from teacher to teacher leader.

The true power of the learning community is that the ideas shared are generative. In this model, that generative process included a pay-it-forward concept of growing teacher leaders who supported newer arrivals in the community, who in turn grew into teacher leaders. As we share, as we challenge ourselves to grow as teachers, we set in motion ideas that have the potential to live beyond us.

......

We would like to thank Richard Kitchen, University of New Mexico, and Sherrin B. Hersch (retired) and Despina Stylianou, The City College of the City University of New York, for the advice, support, and encouragement they offered on the writing and revising of this chapter.

REFERENCES

Alexander, Robin J. *Towards Dialogic Teaching: Rethinking Classroom Talk*. Cambridge, UK: Dialogos, 2004.

Ball, Deborah, Mark M. Thames, and Geoffrey Phelps. "Content Knowledge for Teaching: What Makes It Special?" *Journal of Teacher Education* 59, no. 5 (November/December 2008): 389–407.

Chapin, Suzanne H., and Catherine O'Connor. "Academically Productive Talk: Supporting Students' Learning in Mathematics." In *The Learning of Mathematics*, 2007 Yearbook of the National Council of Teachers of Mathematics (NCTM), edited by W. Gary Martin and Marilyn E. Strutchens. Reston, Va.: NCTM, 2007.

Choppin, Jeffrey. "Engaging Students in Collaborative Discussions: Developing Teachers' Expertise." In *The Learning of Mathematics*, 2007 Yearbook of the National Council of Teachers of Mathematics (NCTM), edited by W. Gary Martin and Marilyn E. Strutchens. Reston, Va.: NCTM, 2007.

Cobb, Paul, Terry Wood, and Erna Yackel. "Discourse, Mathematical Thinking, and Classroom Practice." In *Contexts for Learning: Sociocultural Dynamics in Children's Development*, edited by Ellice A. Forman, Norris Minick, and C. Addison Stone, pp. 91–119. New York: Oxford University Press, 1993.

Costa, Arthur L., and Bena Kallick. *Discovering and Exploring Habits of Mind*. Alexandria, Va.: ASCD, 2000.

Darling-Hammond, Linda, and Nikole Richardson. "Teacher Learning: What Matters?" *Educational Leadership* 66 (February 2009): 46–53.

Davis, Brent and Elaine Simmt. "Understanding Learning Systems: Mathematics Education and Complexity Science." *Journal for Research in Mathematics Education* 32, no. 2 (2003): 137–67.

Fosnot, Catherine T., and Maarten Dolk. *Young Mathematicians at Work: Constructing Multiplication and Division*. Portsmouth, N.H.: Heinemann, 2001.

Herbel-Eisenmann, Beth, and Mary Cirillo. *Promoting Purposeful Discourse*. Reston, Va.: NCTM, 2009.

Jensen, Miki, and Catherine Fosnot. "The Box Factory: Extending Multiplication with the Array." In *Context for Learning Mathematics*. Portsmouth, N.H.: Heinemann, 2007.

Lampert, Maggie, and Deborah Ball. "Aligning Teacher Education with Contemporary K–12 Reform Visions." In *Teaching as the Learning Profession: Handbook of Policy and Practice*, edited by Linda Darling-Hammond and Gary Sykes, pp. 33–53. San Francisco: Jossey-Bass, 1999.

Lieberman, Anne, and Lynne Miller, eds. *Teachers in Professional Communities: Improving Teaching and Learning*. New York: Teachers' College Press, 2008.

Little, Judith Warren. "Organizing Schools for Teacher Learning." In *Teaching as the Learning Profession: Handbook of Policy and Practice*, edited by Linda Darling-Hammond and Gary Sykes, pp. 233–62. San Francisco: Jossey-Bass, 1999.

Louis, Karen, S., and Sharon D. Kruse, eds. *Professionalism and Community: Perspectives on Reforming Urban Schools*. Thousand Oaks, Calif.: Corwin Press, 1995.

Reid, David, and Vicki Zack. "Aspects of Teaching Proving in Upper Elementary School." In *Teaching and Learning Proof Across the Grades, A K-16 Perspective*, edited by Despina A. Stylianou, Maria L. Blanton, and Eric J. Knuth, pp. 133–46. New York: Routledge, 2009.

Sherin, Miriam, Vicki Jacobs, and Randy Philipp, eds. *Mathematics Teacher Noticing: Seeing through Teachers' Eyes*. New York: Routledge, 2011.

Shifter, Deborah. "Representation-based Proof in the Elementary Grades." In *Teaching and Learning Proof Across the Grades, A K–16 Perspective*, edited by Despina A. Stylianou, Maria L. Blanton, and Eric J. Knuth, pp. 71–86. New York: Routledge, 2009.

Stein, Mary K., Edward Silver, and Maggie Smith. "Mathematics Reform and Teacher Development: A Community of Practice Perspective." In *Thinking Practices in Mathematics and Science Learning*, edited by James G. Greeno and Shelley V. Goldman, pp. 17–52. Mahwah, N.J.: Lawrence Erlbaum Associates, 1998.

Wenger, Etienne. *Communities of Practice: Learning, Meaning, and Identity.* New York: Cambridge University Press, 1999.

West, Lucy, and Fritz Staub. *Content-Focused Coaching: Transforming Mathematics Lessons.* Portsmouth, N.H.: Heinemann, 2003.

West, Lucy, and Antonia Cameron. *Content-Focused Coaching: Transforming Teacher Practice.* Portsmouth, N.H.: Heinemann, in press.

York-Barr, Jennifer, and Karen Duke. "What Do We Know about Teacher Leadership? Findings from Two Decades of Scholarship." *Review of Educational Research* 74 (September 2004): 255–317.

Part V

Working with Multiple Partners

How can professionals from different settings and roles work together to benefit all?

Our work as educators depends on effective partnerships with a variety of professionals. We continuously work to foster relationships that work—impacting what happens in multiple educational settings. Collaborations require a commitment from stakeholders and must arise out of a shared purpose. Such collaborations involve multiple players who bring a myriad of experiences as well as a variety of personalities to the arena.

Numerous obstacles stand in the way of effective collaborations, including institutional, political, professional, and personal practices, expectations, and beliefs; however, the results of improved practice and the positive impact on students' experiences are invaluable. Successful collaborations require focused attention and concentrated efforts, which can only be sustained through deliberate attention to both formal and informal activities.

The chapters in this section describe collaborations in ways that provide insights into the process of building and sustaining effective professional relationships. As you read them, consider the following questions and try to identify those salient features and practices of professional collaborations that have demonstrated success.

- How do professionals build a collaborative team approach to address complex educational issues?
- In what ways do partners address challenges so that partnerships evolve over time and promote change?
- What are some characteristics of an effective partnership, and how are those characteristics developed and nurtured?

CHAPTER 14

Collaborative Problem Solving as a Basis for a Professional Learning Community

Four Perspectives on Collaboration in the BIFOCAL Project

Edward Silver
Valerie Mills
Dana Gosen
Geraldine Devine

THE BIFOCAL (Beyond Implementation: Focusing on Challenge and Learning) project, a multiyear endeavor, was part of the National Science Foundation–funded Center for Proficiency in Teaching Mathematics (CPTM). BIFOCAL involved one faculty member and several graduate students from the University of Michigan, one mathematics curriculum and professional development specialist from Oakland (Michigan) Schools, and teachers of middle-grades mathematics from five different school districts. In this chapter we share some perspectives on our individual and collective engagement with this project and what we think we gained from it. In doing so, we hope to offer some lessons that can be extracted from our experiences. In this limited space, we omit many details about the BIFOCAL project. More information can be found in the several papers that were produced during the project; these are cited at appropriate places in the chapter and listed among the references.

Gerri's Perspective: A Teacher(-Leader)

When the BIFOCAL project began, I had taught high school mathematics for nine years, and I was transitioning to a new role in which I would both teach some classes and spend part of my time supporting colleagues to use recently adopted, standards-based curriculum materials at the middle school. Though I felt secure in my mathematical content knowledge,

I had not yet devoted concentrated effort to examining my instructional practice, and I had little experience in supporting colleagues to do that. I fully intended to implement the new curriculum with fidelity, and I did not really anticipate any problems doing so. My experiences in BIFOCAL opened my eyes to instructional issues I had not adequately considered, some of which are highlighted below. In BIFOCAL, the use of mathematics tasks, cases, and modified lesson study helped illuminate aspects of instructional practice that were relevant to me. Using these tools and processes, I could envision new pedagogy and learn how to analyze my practice. With each of the cases, we were asked to give specific evidence in the case. Stating evidence from the cases helped me consider cause-and-effect relations. That is, I explored details about tasks and teacher moves in enacting tasks, and I looked for related evidence of student understanding. For example, as an avid user of choral response I was shocked to see how little evidence I could find about student understanding involving choral response. This revelation challenged me to investigate and interrogate the value of other teaching techniques and to integrate new strategies in my classroom. Because the project was comprised of multiple, connected sessions, I could continually integrate, reflect upon, and modify instructional strategies with additional insights from each session.

Early on, I was consumed in the work of improving my classroom practice and did not give much attention to my leadership practice. However, as the project continued, I moved from examining the interplay of teachers, students, and mathematical content—as depicted in the cases—to examining the interaction of teacher-leaders, teachers, and instructional content, as this played out in the BIFOCAL sessions. One pivotal moment in this transition was when I realized that the questions I needed to ask myself when planning for professional development were complementary to those we were using to plan classroom lessons collectively. Key to this transition were the regular planning team conversations that made visible the origins of the facilitators' probing questions. For example, one question used persistently during professional development sessions was, "What did/will you see or hear that let/(s) you know whether all students in the class understand the mathematical ideas that you intended them to learn?" This question came from the same Thinking Through a Lesson Protocol (TTLP) (Smith, Bill, and Hughes 2008) that we used to plan classroom lessons.

My increased awareness of these features and my transition to leadership was fostered not only by the long-term nature of the project, which allowed me time to implement and refine new learning, but also by the opportunity to join the planning team. Ed's reflection below identifies some key features of the planning sessions. Although I was a novice to planning professional development and using scholarly research, the team welcomed both my contributions and questions. Through extended off-site conversations and collaboration with my district colleagues, I brought to the planning team additional insight into the participants' thinking as well as authentic student work. Building on my experience stating evidence in cases during project activities, I learned from the team how to similarly seek evidence of participant knowledge from session reflections and artifacts. The team then used this evidence to subsequently adapt upcoming sessions (e.g., Silver et al. [2008]).

Now that I have transitioned into a new position as a professional development specialist, I regularly invite other teachers to participate in planning new projects, and I help them to take up evidence and plan accordingly.

The BIFOCAL Project

BIFOCAL was designed to support teachers who were users of standards-based middle school mathematics curriculum materials. In particular, we worked with users of *Connected Mathematics* (Lappan et al. 1996). These materials provide a rich source of worthwhile, complex mathematics tasks for use in instruction, but teachers often face challenges in using such tasks effectively to provide powerful learning opportunities for their students. BIFOCAL began as an effort aimed at solving an instructional challenge that we named the *curriculum implementation plateau*. This is a term we used for the apparent lull that occurred a few years after adopting a set of innovative curriculum materials. Both observation and data suggested that the kind of curriculum-based professional development provided when new materials are adopted was not sufficient to sustain long-term development of instructional effectiveness and student learning. In this sense, it appeared that teachers might have reached a curriculum implementation plateau.

BIFOCAL rested on a foundation of prior work in the QUASAR (Quantitative Understanding: Amplifying Student Achievement & Reasoning) and COMET (Cases of Mathematics Instruction to Enhance Teaching) projects. It was particularly based on research on instruction using cognitively demanding tasks (e.g., Stein, Grover, and Henningsen [1996]; Henningsen and Stein [1997]) and narrative instructional cases designed to draw attention to issues and challenges faced by teachers in using such tasks (e.g., Stein et al. [2000, 2009]). BIFOCAL sessions engaged participants in two interrelated activity strands: (1) the analysis and discussion of an instructional artifact or vignette (e.g., a narrative case or collection of student work); and (2) a modified version of lesson study that included joint lesson planning and analysis. The integration of these strands is described more fully in Silver et al. (2006).

The Mathematical Tasks Framework (MTF) (Stein et al. 2000, 2009) was used as a core framework in BIFOCAL. The MTF (see fig. 14.1) portrays the passage of instructional tasks as they move from the pages of curriculum materials into the work of teachers and students in classrooms, with special attention to the modifications that teachers may make, intentionally or unintentionally, to reduce the cognitive complexity of challenging tasks. The MTF thus draws attention to the importance of a teacher's actions and interactions in supporting students' engagement with cognitively demanding tasks. In addition, the MTF points to the critical importance of both the cognitive demand of tasks and the role teachers play in influencing the opportunities that students have to learn mathematics through engagement with complex tasks. We also used a set of narrative cases closely tied to the MTF (e.g., Smith, Silver, and Stein [2005]) and some associated auxiliary tools (e.g., Thinking Through a Lesson Plan [TTLP] Protocol). For more on how the framework guided the work of the project, see Valerie's commentary later in the chapter.

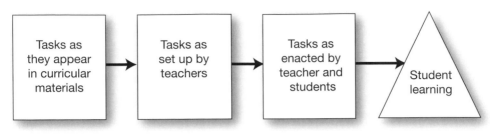

Fig. 14.1. Mathematical Tasks Framework (MTF)

Ed's Perspective: A Faculty Member

I have had many opportunities over the years to work with others in joint ventures. My experience in BIFOCAL was distinctive in ways that brought forth some key issues related to the possibilities for productive interplay between research and practice in mathematics education.

Education research is commonly criticized for not being relevant to educational practice, and at the same time for being driven more by opinion and fad than by solid scientific research and evidence. In general, the disconnection between research and practice in education (and mathematics education is no exception) is accepted as a fact of life. Yet, when I reflect on the work we did in BIFOCAL, I see many examples of dynamic, interactive interplay between research and practice. Moreover, I think the three features identified in the next section of this chapter were key reasons why the research-practice interplay was so central to the work of the project.

One example of the research-practice interplay in BIFOCAL was our decision to use research-based knowledge and resources to address an issue of practice; namely, we used the MTF and MTF-based cases as core elements of our professional development response to the implementation plateau. But that was just a part of the research-practice interaction in the project. We also engaged in careful documentation of project sessions with teacher participants, and then used that documentation as a source of further research and learning. We videotaped sessions, collected artifacts of participants' work, and required end-of-day reflections, and we then examined these data regularly not only to ascertain "how things were progressing," but also to refresh and revise our understanding of the problem of practice at the heart of the project, and the ways in which our professional development was (or was not) addressing the needs of participants.

We accepted from the beginning that we had an incomplete understanding of both the problem and the best way to address it in professional development, so we were able to take a skeptical stance toward our work. We believed we were doing something that could be helpful to the teacher participants, but we did not believe it so fervently that we would not persistently strive to improve both our understanding of the core problem and our approach to addressing it. In this regard, having a heterogeneous group of collaborators was wonderful, because challenges to a current state of understanding could (and often did) come from different perspectives. A primary site for much of our collaboration and joint learning was our planning meetings, as Gerri noted earlier in her reflection. These meetings were held regularly throughout the year, often biweekly. In the meetings, we engaged in data-based debriefing of prior sessions, which in turn led to decisions about the next facet of the problem that warranted attention and the best way to address that facet in future sessions. We discussed and debated our individual perceptions, our priorities for future sessions, and our choice of tactics and resources for the next professional development session. Our discussion and debate was grounded very much in the data that had been collected and analyzed along the way. Thus, as Dana explains in her reflection below, project collaborators used research (i.e., inquiry-driven data collection and analysis) to design a professional development practice that helped to address an authentic problem of teaching practice and also to provide a setting in which we could generate new knowledge about the problem.

In this way, and others, BIFOCAL served not only as a provider of professional development but also as an incubator for the interaction of research and practice. BIFOCAL generated three years of professional development sessions for project participants, several professional presentations and published papers in which the collaborative team identified issues relevant to the practice of mathematics teaching and mathematics teacher professional development, and transformative experiences for all of us who participated in the learning community created by the project (e.g., Silver et al. [2007]).

Collaboration in BIFOCAL: Some Key Features

In our view, BIFOCAL was truly a collaborative group effort. To explain what we mean by that term, we find it useful to refer to a distinction offered by Roschelle and Teasley (1995) between cooperation and collaboration in regard to group functioning and group learning. Working and learning in cooperative groups may be said to rely on the effective distribution of responsibilities among participants with different knowledge and expertise (Slavin 1995). In contrast, Roschelle and Teasley suggest that collaborative working in groups, and learning in collaborative groups, relies on continuous, joint effort that blends different forms of knowledge and expertise.

We do not argue that collaborative group work is necessarily superior to cooperative group work; in fact, in many ways it is likely more difficult both to establish and to sustain. Nevertheless, we use this distinction to draw attention to our first key aspect of BIFOCAL; namely, that it was *a sustained effort over time in which individuals with different spheres of knowledge and expertise engaged together in joint activity aimed at a commonly held problem.* In BIFOCAL the commonly held problem was the so-called curriculum implementation plateau. The varied spheres of expertise held by the team included research and scholarship on mathematics teaching and learning; extensive knowledge and experience in identifying strategic areas of need and in designing and delivering effective professional development for teachers; and firsthand knowledge of and experience in using the curriculum materials, including knowledge of the embedded enactment challenges. The project team also shared a strong inclination not only to design and deliver professional development but also to study seriously both its enactment and its effects.

Closely related to the first feature is another key feature of the collaboration in BIFOCAL; namely *the expertise possessed by each partner in the collaboration was respected and treated as legitimate in the joint work, and the partners joined together with an intention to learn from each other in and through the collaboration.* One of the reasons we formed the BIFOCAL collaboration was the recognition that none of the partners possessed the required knowledge and skill to address the core problem, but each partner had expertise others might not that was a likely component of a successful effort to address the problem. Moreover, we were willing to venture beyond the boundaries of our traditional roles in research or practice to learn from and with one another. For example, those with a strong scholarly interest were willing to help define and address a problem of practice. They saw in this work an opportunity to learn about both the practical issues entailed in designing and enacting professional development for experienced teachers and as well as about the nature of teachers' work in using innovative mathematics curriculum materials. Those with a strong practice-oriented interest, on the other hand, seized the opportunity to learn how to study their work and make it an object of reflection for improvement. Thus, we came together as a professional learning community to engage in this collaborative work.

A third key feature of the collaboration was *a problem-solving orientation that allowed for attention to and continuous refinement of our understanding of both the problem and how we might address it.* We began with a vaguely defined problem: after several years of using *Connected Mathematics*, teachers and students did not appear to be progressing to higher levels of implementation and achievement. This concern was based on anecdotes and a few suggestive tidbits of data, but we certainly did not have a solid understanding of the curriculum implementation plateau at the outset. We also began with a hunch that something we knew from prior research on teachers' use of cognitively demanding tasks in the mathematics classroom might be implicated, but we certainly did not know how or why. Our understanding of the problem and how we might address it was refined over time, as we interacted with the teacher participants on issues of cognitive demand and came to understand in greater depth how and why certain aspects of teaching with innovative curriculum materials remained persistent challenges (e.g., Silver et al. [2005]).

Dana's Perspective: A Graduate Student

"Transformative" is the word that Thompson and Zeuli (1999) use to describe the kind of learning needed to support "changes in deeply held beliefs, knowledge, and habits of practice" (p. 342). I believe that my participation in BIFOCAL enabled me to learn in transformative ways at the intersection of research and practice (see Ed's earlier reflection). I entered the group as a graduate student with little research experience; seven years later, I am an experienced professional learning specialist and a recent graduate with a PhD in education.

In BIFOCAL we worked individually and collectively on what Lampert (2001) would call a "problem of (teaching) practice." We worked on a common problem, though we differed in our reasons for being interested in this problem. For some BIFOCAL colleagues, the problem was interesting for scholarly reasons; it was an opportunity to understand, examine, and analyze the implementation plateau. For me, a novice graduate student in an unfamiliar world, fiercely holding on to my identity as a middle school mathematics teacher, my interest at the time was far less academic—BIFOCAL was simply a research project that made sense. The project was an opportunity to help mathematics teachers learn to teach mathematics better. Though I was a novice with regard to research, my teaching experiences and expertise afforded me legitimate access to the problem of practice that was the focus of our collective work.

Having access to the problem made my participation in the project productive. Because the problem directly related to teaching practice, it was feasible to connect my new experiences to what I understood about teaching mathematics. As such, the focus of the group's work was not beyond my intellectual reach. While I was not familiar with the scholarship on the effective design of professional learning, I had experience as a teacher and a participant in professional learning settings. Thus, I could offer ideas based on my practical experiences, as opposed to my limited knowledge of the field. Furthermore, I was concurrently engaged in scholarship focused on teacher education research, including professional learning, in my graduate coursework. Together, my work on BIFOCAL and my graduate studies supported my learning as I transitioned from graduate student to teacher educator.

My participation in BIFOCAL was an apprenticeship that integrated research and practice. Through participating in project planning meetings and the professional learning sessions, I created a vision and understanding of effective professional learning grounded in research-based methods and tools. Two of these tools were the MTF and TTLP. These complementary tools were flexible and adaptable enough to be used as foundational elements in the design of each BIFOCAL session. Those characteristics also enable my colleagues and me to continue to integrate them in every K–12 professional learning project we undertake. We have found these tools provide a common language and a coherent thread that teachers, in different courses and different settings, can use over time to scaffold a deepening understanding of the complex ideas associated with the work of teaching, especially those related to selecting, setting up, and enacting mathematical tasks. Some evidence suggests these complementary tools can act to scaffold the development of teacher leadership practices (Gosen 2010).

To reiterate, the nature of the problem explored in BIFOCAL was critical to my learning. It was a problem practical enough to be accessible upon my initiation into the group. Yet, it was also rich enough to sustain scholarly work for several years. Had it been outside my intellectual reach or less intellectually rich, I (as well as others on the team) may not have been able to make the same strides in our learning over the course of the project. Throughout the project, I learned the importance of engaging teachers in mathematical tasks, the benefits of focusing professional learning on a core set of instructional goals and research-based tools, and useful methods for facilitation. I had ample time and many opportunities to refine my understandings and skills. I began participating in the professional learning sessions by planning and observing. Later, I began leading small- and large-group conversations. As time went on, I learned to use education research and evidence from sessions to support my ideas, rather than relying only on past personal experiences to make claims or suggestions. In these and other ways I transformed my thinking with respect to professional learning.

As the previous paragraphs attest, my experience in BIFOCAL was an apprenticeship at the boundary between research and practice. I learned through interactions with people similar to and different from me, people with varied expertise and experience. I found that participating in an integrated apprenticeship of research and practice was a productive experience for me, and it might be for others as well, especially if the apprenticeship can be grounded in an intellectually rich and authentic problem of teaching practice.

Valerie's Reflection: A Professional Learning Specialist

I've worked with teams on various educational projects throughout my career. In earlier projects I gained insights and expertise from colleagues as we shared project work based on personal skill sets, but BIFOCAL was a distinctly different working and learning environment. This project challenged partners to work in ways that were not typical of their jobs and on tasks that required more than individual expertise. The key to the BIFOCAL story with respect to my growth lies in what we did and did not do in framing and enacting our approach to the implementation plateau.

When BIFOCAL began, the implementation plateau was a new phenomenon that was not well understood. As Ed discussed earlier, our team chose to work with the ambiguity inherent in this problem. This approach allowed us to organize our work so that we could develop a better understanding of the implementation plateau and at the same time seek to remediate its effects. Without a solid understanding of the problem, I could not rely on my standard approach to professional learning, which was looking within existing education research for direction. Instead, the project gave me the opportunity to expand my consumer relationship with education research and adopt some of the tools and strategies of a researcher.

The need to adjust my practice provided me with insights that have had a lasting impact on my practice. One example was that I discovered the benefits of greater data collection and analysis than is typical in my work. Our collaborative and exploratory approach to BIFOCAL research required me to regularly collect and analyze data in preparation for the design of each professional learning activity. As a project team we chose not to simply divide the work with university faculty and graduate students taking responsibility for data collection and analysis, and myself, as the professional learning facilitator, planning the next session. Instead we worked together to decide what data would be collected (such as questionnaires, end-of-day reflections, and video transcripts) and how it would be analyzed. Our analysis following each session was then used to craft the next session of professional learning. The fact that we did not choose to partition the project work meant that the research activities were fully informed by the instructional activities and vice versa. This experience has shifted my thinking and practice so that data collection and analysis are now an intrinsic part of my work. Today, my colleagues and I videotape and transcribe professional learning sessions, collect classroom observation data, preserve end-of-day reflections from sessions, participate in end-of-session facilitator debriefs, and archive session posters and teacher/student work electronically. Each element is analyzed for evidence of teacher thinking or practice as they relate to the goals of teachers and the professional learning series. This data is then used to inform the design of future sessions in a manner that is both iterative and adaptive, as was true in BIFOCAL (Silver et al. 2008). We view our professional activity as both informed by and designed as education research.

A sample of our current use of data can be seen in the approach we take to the design of professional learning. The development of each session agenda begins with an examination and analysis of the prior session's data. This evidence of the current thinking of participating teachers, and their stated needs or concerns, is considered in comparison to our series goals. We then use this information to suggest next learning steps for participants and to inform decisions such as the selection of discussion questions, readings, and tasks in the next session. For example, we regularly use a summary of the end-of-day reflection data in which we share participants' current thinking in their own words to launch the next session in a series. A discussion of the compiled reflections is followed by a journal prompt asking participants to consider the comments of their colleagues. These activities help to connect sessions separated by time, highlight series goals, inform participants about the ways in which their colleagues' thinking is evolving, and explicitly connect ideas discussed in the sessions to classroom practice. The data we collect informs both the design of future sessions and the thinking of participants as they explore new ideas.

I have worked on the practitioner side of the research-practice boundary for most of my career, but BIFOCAL offered me an opportunity to move with colleagues across the boundary in ways that blurred the distinction between research and practice. As a result, my thinking about my professional learning practice is richer and I believe more effective. Being good at what we do on either side of the practice-research divide takes years of focused effort and leaves little time to expend energy beyond the constraints of our job description. However, this experience has left me wondering what might be gained on both sides of the divide if, through collaboration on commonly held concerns, each understood more about the other's work.

Closing Remarks

We believe that the individual and collective insights and growth that emerged for us in BIFOCAL were due in no small part to our decision early on to tackle the curriculum implementation plateau, despite its complexity and ambiguity. This leads us to one final lesson that we think can be drawn from our work: To create communities of practice in professional learning initiatives, we urge a collaborative team approach, in which the partners take a research-based and problem-solving-oriented stance, seasoned with an ample dose of skepticism, toward an authentic problem of practice, and then work on a long-term basis to address it. Based on our experience in BIFOCAL, we posit that doing so will foster opportunities for learning and growth that are powerful and lasting for all members of the collaboration.

......

This article is based upon work supported in part by the National Science Foundation under grant number 0119790 to the Center for Proficiency in Teaching Mathematics and in part by the Michigan State University Mathematics Education Endowment Fund for support of the BIFOCAL project. The Usable Scholarship Initiative at the University of Michigan provided support for the preparation of the paper. Any opinions, findings, and conclusions or recommendations expressed in this material are those of the authors and do not necessarily reflect the views of the NSF, the Center, or the university. We are grateful to the other teachers and graduate students who were part of the learning community in the BIFOCAL project; we learned so much from this collaboration.

REFERENCES

Gosen, Dana. "A Case Study Analysis of the Role and Influence of Tools in the Practice of a Teacher(-Leader) within the Context of a Professional Development Project." Unpublished doctoral dissertation, University of Michigan, Ann Arbor, 2010.

Henningsen, Marjorie, and Mary Kay Stein. "Mathematical Tasks and Student Cognition: Classroom-based Factors That Support and Inhibit High-level Mathematical Thinking and Reasoning." *Journal for Research in Mathematics Education* 28, no. 5 (1997): 524–49.

Lappan, Glenda, James T. Fey, William M. Fitzgerald, Susan N. Friel, and Elizabeth E. Phillips. *Connected Mathematics Project*. Palo Alto, Calif.: Dale Seymour, 1996.

Lampert, Magdalene. *Teaching Problems and the Problems of Teaching*. New Haven, Conn.: Yale University Press, 2001.

Roschelle, Jeremy, and Stephanie Teasley. "The Construction of Shared Knowledge in Collaborative Problem Solving." In *Computer-Supported Collaborative Learning*, edited by Claire E. O'Malley, pp. 69–97. Heidelberg, Germany: Springer-Verlag, 1995.

Silver, Edward A., Lawrence M. Clark, Hala Ghousseini, Charalambos Charlambous, and Jenny T. Sealy. "Where Is the Mathematics? Examining Teachers' Mathematical Learning Opportunities in Practice-based Professional Learning Tasks." *Journal of Mathematics Teacher Education* 10 nos. 4–6 (2007): 261–77.

Silver, Edward A., Hala Ghousseini, Dana Gosen, Charalambos Charalambous, and Beatriz T. F. Strawhun. "Moving from Rhetoric to Praxis: Issues Faced by Teachers in Having Students Consider Multiple Solutions for Problems in the Mathematics Classroom." *Journal of Mathematical Behavior* 24, nos 3–4 (2005): 287–301.

Silver, Edward A., Valerie Mills, Hala Ghousseini, and Charalambos Charalambous. "Exploring the Curriculum Implementation Plateau: Understanding and Confronting Issues and Challenges." In *Mathematics Teachers at Work: Connecting Curriculum Materials and Classroom Instruction*, edited by Janine Remillard, Beth Herbel-Eisenmann, and Gwendolyn Lloyd, pp. 245–65. London: Routledge, 2009.

Silver, Edward A., Valerie Mills, Alison Castro, and Hala Ghousseini. "Blending Elements of Lesson Study with Case Analysis and Discussion: A Promising Professional Development Synergy." In *The Work of Mathematics Teacher Educators: Continuing the Conversation*, edited by Kathleen Lynch-Davis and Robin L. Ryder, pp. 117–32. San Diego: Association of Mathematics Teacher Educators, 2006.

Silver, Edward A., Lawrence Clark, Dana Gosen, and Valerie Mills. "Using Narrative Cases in Mathematics Teacher Professional Development: Strategic Selection and Facilitation Issues." In *Cases in Mathematics Teacher Education: Tools for Developing Knowledge Needed for Teaching*, edited by Margaret S. Smith and Susan Friel, pp. 89–102. San Diego: Association of Mathematics Teacher Educators, 2008.

Slavin, Robert E. *Cooperative Learning: Theory, Research, and Practice.* 2nd ed. Boston: Allyn & Bacon, 1995.

Smith, Margaret S., Victoria Bill, and Elizabeth K. Hughes, "Thinking through a Lesson Protocol: A Key for Successfully Implementing High-level Tasks." *Mathematics Teaching in the Middle School* 14 (2008): 132–38.

Smith, Margaret Schwan, Edward A. Silver, and Mary Kay Stein. *Improving Instruction in Rational Numbers and Proportionality: Using Cases to Transform Mathematics Teaching and Learning, Volume 1.* New York: Teachers College Press, 2005.

Stein, Mary Kay, Barbara Grover, and Marjorie Henningsen. "Building Student Capacity for Mathematical Thinking and Reasoning: An Analysis of Mathematical Tasks Used in Reform Classrooms." *American Educational Research Journal* 33, no. 2 (1996): 455–88.

Stein, Mary Kay, Margaret S. Smith, Marjorie Henningsen, and Edward A. Silver. *Implementing Standards-based Mathematics Instruction: A Casebook for Professional Development.* New York: Teachers College Press, 2000.

————. *Implementing Standards-based Mathematics Instruction: A Casebook for Professional Development, Second Edition.* New York: Teachers College Press, 2009.

Thompson, Charles L., and John S. Zeuli. "The Frame and the Tapestry: Standards-based Reform and Professional Development." In *Teaching as the Learning Profession: Handbook of Policy and Practice,* edited by Linda Darling-Hammond and Gary Sykes, pp. 341–75. San Francisco: Jossey-Bass, 1999.

CHAPTER 15

Sustaining Effective Professional Development School Relationships

Jackie Menser
Adam Harbaugh
Claudia Cox
David K. Pugalee

THE IMPORTANT WORK OF IMPROVING mathematics education will not be realized without focused commitments to links between efforts at the university level and in the schools. Developing and sustaining positive relationships between universities and schools requires time, and diligent effort by both parties. Such partnerships take many forms and have varying degrees of involvement from multiple stakeholders. One form of partnership is the professional development school, or PDS. Since the 1980s, the PDS label has referred to unique and intense school and university collaborations designed to accomplish four goals: preparing future educators, providing current educators with ongoing professional development, encouraging joint school–university faculty investigation of education-related issues, and promoting the learning of P–12 students (National Association for Professional Development Schools [NAPDS] 2008).

This chapter will describe a PDS relationship between an urban university and an urban middle school. Randolph Middle School serves approximately 1,000 children each year, and it houses an International Baccalaureate magnet school. An average of 87 percent of students performed at or above grade level in reading and math according to recent assessment data. The student population is among the most diverse in the city, with more than thirty-two different nationalities represented and with nearly 75 percent of students belonging to a minority group. The University of North Carolina at Charlotte is North Carolina's urban research institution with an enrollment of more than 25,000 students. Its College of Education enrolls more than 3,000 students in undergraduate and graduate professional education programs and recommends more than 600 students each year for a teaching license. (This PDS relationship involves multiple university faculty and school personnel in a multicontent collaboration; this chapter will focus on the mathematics component of that program.)

Our purpose here is not to embark on a program evaluation of the partnership, but to present a useful discussion that will explicate both the benefits and the challenges of this collaboration. Perspectives from both the school and the university will be presented. The authors of this chapter include the principal and the academic facilitator of the middle school as well as two mathematics education faculty members involved in the partnership. The school principal has a background in mathematics, and the academic facilitator has a background in literacy. The two university faculty members specialize in middle and secondary mathematics education. Before exploring the details of this partnership, we will present a short background on research related to PDSs, followed by descriptions of the partners. The next sections will discuss the partnership from the school and university perspectives. A description of an effective project is then provided to illustrate the impact of the partnership. We conclude with some recommendations that have emerged from our experiences.

Defining Professional Development Schools

One difficulty in describing a PDS relationship is the wide array of programs that use the label. In recognition of this identity problem, the National Association for Professional Development Schools (2008) identified nine essential components (see fig. 15.1). Note that these are *essential* components. These nine elements establish the foundation for a PDS and provide a clear set of standards for defining such a partnership. This chapter focuses on four of these nine components: (2) a culture of active engagement, (3) needs-based professional development, (5) shared investigations of practice, and (6) clearly defined roles and responsibilities.

Research indicates that PDSs have positive effects on both preservice and in-service teachers. Professional development schools provide the types of experiences that assist aspiring teachers in developing the set of skills and knowledge necessary for successful induction into the profession. In a longitudinal study comparing student teaching graduates from professional PDSs and those from traditional programs, Latham and Vogt (2007) found that even when controlling for student background and cognitive characteristics, a PDS experience promotes graduates' entry into and perseverance in teaching. These teachers likely had greater efficacy and a sense of empowerment (Ruscoe et. al 1989). Castle, Fox, and Souder (2007) found that PDS teacher candidates scored significantly higher than their non-PDS peers on the characteristics of assessment, instruction, planning, and classroom management. They also claim that PDS teacher candidates may positively affect student achievement sooner than non-PDS candidates. The positive impact on student learning that results from focused efforts at PDSs has also been documented (Castle, Arends, and Rockwood 2008; Bay-Williams, Scott, and Hancock 2007; Creasy 2005; Teitel 2004; Lefever-Davis and Heller 2003).

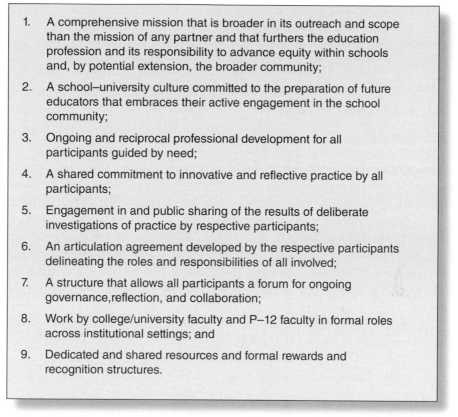

1. A comprehensive mission that is broader in its outreach and scope than the mission of any partner and that furthers the education profession and its responsibility to advance equity within schools and, by potential extension, the broader community;

2. A school–university culture committed to the preparation of future educators that embraces their active engagement in the school community;

3. Ongoing and reciprocal professional development for all participants guided by need;

4. A shared commitment to innovative and reflective practice by all participants;

5. Engagement in and public sharing of the results of deliberate investigations of practice by respective participants;

6. An articulation agreement developed by the respective participants delineating the roles and responsibilities of all involved;

7. A structure that allows all participants a forum for ongoing governance, reflection, and collaboration;

8. Work by college/university faculty and P–12 faculty in formal roles across institutional settings; and

9. Dedicated and shared resources and formal rewards and recognition structures.

Fig. 15.1. Nine essential components of a Professional Development School

Professional Development Schools also have a significant impact on the reflective practices and teaching methods of in-service teachers. Darling-Hammond, Bullmaster, and Cobb (1995) concluded from a study of seven professional development schools that the school elicited a greater degree of teacher commitment; 70 percent of teachers in the study reported that they changed the way they thought about their practice, and more than half reported changes in their methods of teaching. The school-university relationship fosters what Bay-Williams, Scott, and Hancock (2007) refer to as "simultaneous renewal" grounded in change at both the school and university level as well as joint ownership of problems and successes.

Perspectives from School Administrators

Although each of the nine components is essential to developing and maintaining a thriving PDS, we have found that four of them—maintaining a culture of active engagement of all stakeholders, a commitment to needs-based professional development, shared investigations of practice, and clearly defined roles and responsibilities—have been crucial in the success of our school and our relationship with our university partners. Below, we discuss each of these four components from the perspective of the school administrator. (The section that follows this one discusses these four components from the university partner perspective.)

A Culture of Active Engagement

From a school administrator perspective, benefits abound in our PDS relationship. From the moment the university students step into our school as observers, interns, or student teachers, they are invited and required to actively participate in the educational process. This participation often begins with observations of and discussions with our teachers. It is through these observations and discussions that the university student is first introduced to the place where theory meets practice.

One of the underlying strengths of the PDS relationship is the perpetuation of effective teaching. To have interns, observers, and student teachers regularly coming into the school building and observing and interacting with our teachers has become part of the culture of the school. Our teachers are not distracted by the potential disruption of observers coming in and out of their classrooms.

One of the most important things we offer the university students is the opportunity to experience an urban setting with a diverse group of students. As they observe instructional concepts they have learned in university courses being practiced, the university students are witnessing and applying these concepts with a diverse group of students, and not just with affluent students or just one racial or religious group. The university students can see that what they are learning in their courses is applicable to a broad range of students. Diversity is an important part of the culture of our school, so bringing in interns, observers, and student teachers helps to show a broader view of quality education than what many individual university students may have experienced in their own middle school education or other field-based clinical experiences. A number of our student teachers have come from smaller or more rural schools, and they have a particular level of anxiety about coming into an urban school setting that is different from the one they attended as a student. By being able to observe and then complete their internship at our school, many university students' viewpoints are broadened.

Inviting our interns to fully participate in the teaching process helps them to go beyond the level of passive observation. Early in our relationship, interns actively participate in what goes on at the school, including helping with tutorials or presenting mini-lessons to students in the classroom, as well as several other important aspects of what our teachers do. This active participation allows the intern to feel as though he or she were a member of the school community rather than an outsider. The teachers have also learned how to invite participation by the interns, and they no longer have to be prepared by a supervisor on how to include an incoming intern in the teaching process. The open and transparent nature of the school community is an important part of the school culture.

One of the more tangible benefits for school administrators in this relationship has been the opportunity to hire student teachers, especially in mathematics and science where there are acute shortages, which is a way to ensure sustainability of the culture we have established.

Needs-based Professional Development

From the perspective of the university partners, this access to newly trained preservice teachers brings current knowledge based on the latest research; this is especially true in areas of integrated curriculum, such as using mathematics to improve literacy skills, projects incorporating best practices, and use of the latest classroom technologies to enhance student engagement and learning. The sharing made possible through the PDS relationship creates an air of excitement about teaching and learning, which helps to broaden the entire faculty's knowledge of particular topics. This sharing of instructional practices also benefits the student teachers and interns. For example, lesson planning and design is an integral component of the instructional process. Interns and student teachers also get specific training related to the magnet International Baccalaureate plan at the school and the use of rubrics as part of the grading process. Technology is an area where many of our teachers benefit greatly by working with university students, particularly those teachers who have been out of school for a long time. University students are expected to effectively incorporate technology into their lessons. This expectation forces the university students to, in a way, be more experienced than the classroom teacher. This role reversal for both the intern and the teacher helps to develop a productive and dialogic working relationship. As a result of a focus on literacy across the curriculum, the university mathematics educators are working with the mathematics teachers in addressing reading and writing in their classes and are leading professional development while also supporting the teachers in implementation of strategies in their instruction. Such "needs-based" professional development is facilitated through an effective PDS partnership.

Shared Investigations of Practice

The partnership has enriched the professional experiences of both the administrative and instructional staff. The staff have had multiple opportunities to be involved as presenters at national conferences, including PDS conferences. This has been a great benefit for them. For example, the principal, academic facilitator, and a teacher joined university faculty in co-presenting the findings from the quantitative reasoning project at the Research Council for Mathematics Learning annual meeting. The academic facilitator, who has a language arts background, co-presented with university faculty on writing in mathematics at a regional NCTM conference. We continue to work on a metaphor project in which students use metaphors to describe their views about mathematics. Student metaphors reveal that students have complex views about mathematics and predominately see mathematics as a journey—not always a positive one, but one characterized by the hard work and perseverance that many see as necessary to be successful. This work has helped us better understand students' beliefs and how they are affected by classroom experiences. Engaging in school-based research provides an opportunity to use data to reflect on our practices.

The PDS relationship provides a forum for collaborating on the design and implementation of investigations centered on instruction and learning. More important, the results provide an opportunity for the entire PDS community from both the school and university to engage in frank discussions about what the results reveal and what challenges or weaknesses might need to be addressed and how. These rich experiences allow us to design investigations, discuss findings, and use that information to inform our practice.

Clearly Defined Roles and Responsibilities

The personal relationships between administrators and teachers at our school and university faculty and staff prevent us from thinking of the university as a big, impersonal organization across town. We understand the various responsibilities that come with the partnership. These relationships demand that we pick up the phone or send an email if we need something, such as when an issue arises with an intern or we need some specific professional development or want some of our teachers to take a particular course. The personal relationship provides a level of trust that has been built up not so much over time but over a number of shared experiences and challenges. With this trust comes knowledge that the university will provide support when we need it. Conversely, university faculty and supervisors know that they can place interns and student teachers at our school for clinical experiences even if those candidates have had difficulties in other settings and that we will be supportive. We believe that through our shared experiences over the last decade our administrators, teachers, and the university faculty and staff have developed a mutual dependability.

Recommendations

From the perspective of the school administration, recommendations based on lessons learned from building the partnership include the following:

1. Be upfront in sharing with your own school staff about what a professional development school partnership is—the history, the benefits, and the ideas of such a relationship. Understanding the relationship assists in developing an understanding about the dynamics of the partnership.

2. Create a "buy-in" staff ambiance by applauding the efforts of the school's teachers and staff who are supportive of the school's role in preservice teaching. It takes the entire school to make the PDS relationship work, and recognizing efforts from all players is critical in communicating the importance of being involved in the partnership.

3. Choose a university carefully. Be conscientious about what is to be accomplished and keep work aligned to the expectations of all parties, with the priority always being the students in your school.

Perspectives from University Partners

The following sections provide some analysis of PDS partnerships from the point of view of the university partners.

A Culture of Active Engagement

One of the most valued benefits to the university and our students of a strong PDS relationship is having a bridge between teacher education from a university perspective and from a K–12 perspective, where K–12 teachers and administrators are also considered to be teacher educators. Practicing teachers and administrators supervise and teach the teacher candidates in important clinical settings including classroom projects, observations, tutoring, and student teaching. This bridge allows for a more seamless alignment of agendas and focuses the efforts of all school personnel, current and prospective, toward helping effectively teach the K–12 students. Another benefit to having various PDS relationships is that we can ensure, with some level of confidence, that teacher candidates will be placed in clinical experiences with effective teachers and school administrators who generally agree with the philosophies of teaching supported by the university and greater educational community. Teachers can be effective teacher educators. Often, with the opportunity and an appropriate amount and type of support, practicing teachers effectively teach and mentor preservice teachers during clinical experiences, including internships and student teaching. Candidates have opportunities to engage in applying knowledge and skills related to practice in a real environment.

Needs-Based Professional Development

Working as partners in school environments informs the work of university faculty in meaningful ways. In essence, the experience provides a foundation that guides our own knowledge and growth as professionals. For example, one of the challenges facing faculty is the struggle between theory and practice; that is, making our teaching relevant based on what happens in classrooms. This has become especially relevant in recent work around assessing and modifying university teacher-education programs—a process referred to as "revisioning." The close contact with schools as part of the PDS relationship during this period of program revisioning positively benefited the university, our students, our programs, and the partnerships with the schools. An open dialogue with our PDS partner about changes to our programs assisted in grounding our changes to courses and our assessment system to what would most positively impact candidates in the field. Sharing this process created a community where serious discussions occurred on how to best prepare candidates to be more effective when they are in their own classrooms. Further, providing the schools with a rationale for particular aspects of program modifications and sharing the university's approach to candidate assessment helped ease tensions that resulted when there was not a shared vision of what we were trying to accomplish.

Shared Investigations of Practice

One of the ways university faculty benefits from being part of this school-level community is through access to teachers and students as participants in research. Faculty have responsibilities as scholars, and it is imperative that they develop PDS relationships that will support their research. This access helps university faculty contribute to the larger learning community through dissemination of research results while informing practice in the schools. Additionally, by having teacher candidates meaningfully involved in the school-level learning community, the university is benefiting by ensuring they are graduating teachers who are prepared to identify, investigate, and work to solve existing and emergent problems within their future classrooms and schools. Shared investigations of practice that are data-driven and derived from sound research practices allow faculty to support the goals of the PDS while building on that work in ways that result in tangible products such as presentations and publications. The illustration of one such successful project is shared later in this chapter.

Clearly Defined Roles and Responsibilities

In order to be an effective learning community, all members need to experience a give-and-take relationship. The partners need to feel as though they are both contributing to the community and getting something in return from their being part of the community. At the school level, an effective PDS can be a learning community where teachers and administrators work to identify problems based on experiences and student data and then work together to solve these problems on a deep level, similar to a lesson study approach. University faculty can be an integral part of this problem-solving process in many ways. They bring an outside perspective and therefore can help identify potential issues in the school, as well as potential solutions grounded in educational research. University faculty can also help to collect and analyze student data in meaningful ways beyond achievement results from state-administered exams. They can also contribute to the community by providing material resources or help in getting funding for these needed resources.

One frustration, from the university perspective, is a weak evaluation system for evaluating the partnerships and sometimes a lack of accountability for the PDS liaisons to the individual schools. In order to be most productive, partnerships require open lines of communication and constant assessment of their rationale and goals, and what the partnership community will be in terms of numbers, types, and involvement of all the stakeholders. Another challenge has arisen in the form of data collection. In the past, the university has collected data from all the PDS schools in a way that made it difficult to answer organizational and research questions about the effectiveness of our programs related to different PDS sites. Data collection has to be part of a comprehensive system and involve preservice teacher measures, in-service teacher measures of instructional performance, *and* student learning outcomes. Recent discussions around program change have resulted in a better alignment between the candidate assessment system and what happens in candidates' experiences in the PDS setting, including the identification of assessments and data that can be used to make decisions about candidate performance and ultimately program efficacy.

Recommendations

Based on the experiences of the university partners, the following recommendations emerge:

1. Provide open communication to the university community about the expectations of being involved with a PDS. Understanding roles and responsibilities provides for a focused and goal-oriented experience while minimizing potential misunderstandings.

2. Stress the benefits of being part of a university-school relationship both in terms of how to align the work with the responsibilities of teaching, research, and service. The work with a PDS should inform all dimensions of the professional responsibilities of faculty.

3. Celebrate accomplishments and honor commitments, while also acknowledging shortcomings. This is central to building a community and trust.

An Illustration of a Successful Project

Each year the partnership team plans a schoolwide focus around an event or topic. One event with a specific mathematics focus revolved around the 2000 census. Every teacher in the school became involved in activities that would support the theme. The event culminated in an extravaganza where students and teachers showcased their projects to parents and guests including district administrators and university faculty. The mathematics component of the project involved students in collecting and analyzing demographic data for their school community. This event, like the other focus events, was successful in that each teacher engaged students so that their understanding was enhanced through connections across subject areas disciplines and through hands-on investigations. For example, one science class collected data on recessive genetic traits (such as tongue rolling and eye color), which allowed them to look at ratios and percentages. (With the guidance of a science educator, students also learned why such "experiments" are an oversimplification of genetics.) A mathematics class developed their own "school census survey," which allowed them to collect, analyze, and present findings on questions related to interests, sibling configurations, countries of origin, academic goals, and more. University partners teamed with classroom teachers and the school support staff to assist teachers with implementing lessons with a census focus.

This particular project was exemplary in that school personnel were involved in extending this work into a school-based research project. The focus of the research was to describe the middle grades students' quantitative literacy skills. This emphasis on quantitative literacy highlighted the increasing importance of data in our lives. This research study provided information about the students' skills and abilities in working with tables, interpreting graphs, and locating and interpreting quantitative information (see Pugalee, Hartman, and Forrester [2008]). Assessments from more than 700 students in grades 6 through 8 showed that students struggle with quantitative literacy skills. More specifically, students struggled with problems that involved making comparisons or multiple steps where results were used in further analysis. As a result of this research component, teachers began to target specific skills related to quantitative literacy as they engaged students in census-related activities in all subject areas.

The principal and instructional facilitator joined university faculty in presenting results of this research at multiple conferences including mathematics conferences and PDS-related conferences. These presentation experiences exemplify the collaborative nature of PDS projects and the shared ownership of the outcomes. Due to the success of the previous project, there are plans for a similar experience around the 2010 census.

Conclusion

The perspectives and information presented in this chapter support the importance of professional development schools to both schools and universities. The National Council for the Accreditation of Teacher Education (NCATE 2001) provided a strong statement in response to this question. Recognizing the critical nature of standards-based reform and school restructuring in addressing the focus on student achievement and accountability, NCATE offered that PDSs are important because they bring together these two streams of reform. They support the necessary alignment, and they go beyond it. "Students, candidates, and faculty benefit as a result of the opportunities to learn in the context of a PDS partnership. PDSs are settings in which new practitioners and P–12 and university faculty can learn to meet the challenges of twenty-first-century education together, because the expertise and resources of both university and the schools support them" (p. 2).

Review the nine essential characteristics of a professional development school offered earlier in this chapter. These characteristics describe a vibrant professional learning community where university faculty and school partners work together to provide opportunities to develop preservice and in-service teacher knowledge and skills with a focus on positively impacting student performance. The experience shared in this chapter offers insights into such a relationship. Partnerships can work, but they require focused attention—successful relationships are not built overnight. The partnership described in this chapter has evolved for over a decade. The perspectives shared here capture both benefits and challenges in making the relationship work and in creating an environment where change can take place. Professional development schools provide a unique opportunity for universities and schools to engage in collaborations that address the challenges to making meaningful change in mathematics education and that have the potential for long-term impact on mathematics teaching and learning.

REFERENCES

Bay-Williams, Jennifer M., Michael B. Scott, and Melisa Hancock. "Case of the Mathematics Team: Implementing a Team Model for Simultaneous Renewal." *Journal of Educational Research* 100, no. 4 (2007): 243–53.

Castle, Sharon, Richard I. Arends, and Kathleen D. Rockwood. "Student Learning in a Professional Development School and a Control School." *The Professional Educator* 32, no. 1 (2008): 1–15.

Castle, Sharon, Rebecca K. Fox, and Kathleen O'Hanlan Souder. "Do Professional Development Schools (PDSs) Make a Difference? A Comparative Study of PDS and Non-PDS Teacher Candidates." *Journal of Teacher Education* 57, no. 1 (2006): 65–80.

Creasy, Kim. "The Effects of a Professional Development School Program on Student Achievement as Measured by the Iowa Test of Basic Skills, Teacher Perceptions of School Climate, and Pre-service Teacher Reflections." PhD diss., University of Akron, 2005.

Darling-Hammond, Linda, Marcy Bullmaster, and Velma Cobb. "Rethinking Teacher Leadership through Professional Development Schools." *The Elementary School Journal* 96, no. 1 (1995): 87–106.

Latham, Nancy I., and W. Paul Vogt. "Do Professional Development Schools Reduce Teacher Attrition? Evidence from a Longitudinal Study of 1000 Graduates." *Journal of Teacher Education* 58, no. 2 (2007): 153–67.

Lefever-Davis, Shirley, and Mary Heller. "Trends in Teacher Certification and Literacy: Literacy Partnerships in Education." *Reading Teacher* 56, no. 6 (2003): 566–68.

National Association for Professional Development Schools. "What It Means to Be a Professional Development School." 2008. http://www.napds.org/nine_essen.html.

National Council for the Accreditation of Teacher Education. *Standards for Professional Development Schools.* 2001. http://www.ncate.org/documents/pdsStandards.pdf.

Pugalee, David K., Kim J. Hartman, and Jennifer H. Forrester. "Performance of Middle Grade Students on Quantitative Tasks: A Beginning Dialogue on Quantitative Literacy in Middle Schools. *Investigations in Mathematics Learning* 1, no. 2 (2008): 35–51.

Ruscoe, Gordon C., Betty L. Whitford, Wynn Egginton, and Mary Esselman. "Quantitative and Qualitative Perspectives on Teacher Attitudes in Professional Development Schools." Paper prepared for the annual meeting of the American Educational Research Association, San Francisco, March 27–31, 1989.

Teitel, Lee. *How Professional Development Schools Make a Difference: A Review of Research.* Rev. 2nd ed. Washington, D.C.: National Council for Accreditation of Teacher Education, 2004.

CHAPTER 16

Teachers Learning through Collaboration

Challenges and Impacts of an Inter-institutional Project

Jacqueline Coomes
Janet Hart Frost
Kristine Lindeblad

T HE DIFFICULTIES STUDENTS FACE as they transition from high school to college mathematics are well known throughout the United States. Nationally, 28 percent of students entering college immediately after high school take at least one remedial mathematics class (Attewell et al. 2006). These students are at a disadvantage for graduating from college, as only 39 percent of students who take remedial classes in college graduate with a four-year degree. In contrast, 69 percent of students who do not take remedial classes graduate with a bachelor's (Attewell et al. 2006). Even students who enroll in college-level mathematics classes face difficulties; failure rates in college algebra are approximately 50 percent across the nation (Gordon 2008). In response to these dire facts, faculty and administrators at high schools and colleges within the greater River City area (all names and places are pseudonyms) formed a professional development project to address the problems of students' preparedness for college mathematics. The project included faculty and administrators from sixteen high schools, two community colleges, and two public universities. The authors of this chapter are project leaders for this project.

In this chapter, we describe how this professional development project led to collaboration within and across institutions and contributed to its participants' shared understanding of high-quality mathematics instruction for all students. First, we describe the participants and structure of the project and the principles that framed the design of the activities. Second, we provide examples of the activities completed by the Professional Learning Communities (PLCs) (Dufour and Eaker 1998). Then we discuss some of the issues and challenges that arose and how we addressed them. Finally, we describe the changes participants made in their practices and how these changes affected student learning. Data and analyses come from focus group interviews, workshop evaluations, online discussions, and case studies conducted by the authors and reported in more detail in Frost and Coomes (2009).

Participants and Structure

An essential question guided the development of the project: *How do we cultivate a community of faculty and teachers who will prepare students for a successful transition from high school to college mathematics?* This question suggested that the problems and solutions were at both secondary and postsecondary levels, because a successful transition implied that students were prepared to enter college-level mathematics and were supported to be successful once they were there. The question also highlighted a need for improved instruction at both levels.

As we designed the project structure and activities, we considered characteristics of successful professional development (PD) and the implications of our essential question. Darling-Hammond et al. (2009) provided four principles for effective PD. Their research results indicated that it should:

1. Be intensive, ongoing, and connected to practice
2. Focus on student learning and address the teaching of specific curriculum content
3. Align with school improvement priorities and goals
4. Build strong working relationships among teachers (pp. 9–11)

We addressed Principle 1—the need for the PD to be intense, ongoing, and connected to practice—in multiple ways. From the beginning of the project, teachers engaged in workshops four times per year and in several activities between workshops. After the first year, we added three components that further supported this principle. The first, an online forum, provided teachers with a place to hold ongoing discussions about teaching and learning, and about the mathematics they teach. Discussing these issues online helped participants to maintain a focus on project goals between the workshops and to relate project goals to their practice. A second added component was classroom observations by project leaders. Each teacher was observed three times per year, between workshops. The observer acted as a coach, meeting with teachers after the observation to discuss what was seen. These observations allowed project leaders to assess how the project was affecting instruction, and to have discussions with each participant that connected the ideas of the project back to the teacher's practice. The third change was to add a three-day summer workshop each year focused on content knowledge.

The second principle of effective PD provides that it must address student learning and the teaching of specific content. The project addressed this principle by including student work in activities at each workshop. Additionally, teachers attended summer institutes focused on mathematics content, and they solved problems in the workshops to improve their understanding of the mathematics they teach. For each of these activities, the mathematical content was aligned with the content from at least one of three sources: our state mathematics standards, the Common Core State Standards (CCSSI 2010), and the College Readiness Standards (CRS) (Transition Math Project [TMP] 2004).

Recognizing that students' difficulties are not only with content, the CRS describe mathematical processes and student attributes necessary for students to be successful in college mathematics. The mathematical processes include reasoning and problem solving, connections, and communication; the student attributes include student behaviors such as engaging intellectually, taking responsibility for learning, paying attention to detail, and persevering. These processes and attributes became more important throughout the project as they provided focal points for many activities and as teachers shared how they incorporated them into their classrooms.

We addressed Principle 4, that effective PD must build strong working relationships between teachers, by designing two group structures: school PLCs (Dufour and Eaker 1998) and the large group of all participants. Each PLC consisted of two to three high school teachers from the same high school and one higher education faculty member. The next two sections provide examples of the work within these two structures.

Large-Group Activities

The large group of all participants met in workshops four times a year, each lasting a day and a half. During the workshops, teachers worked in their school PLCs some of the time, but at other times regrouped to work with participants from other schools. By changing the groups, we cultivated both small- and large-group collaboration and commitment, developing the larger PLC. Many workshop activities focused on improving participants' understandings of instructional strategies that have the potential to increase student learning. To that end, they read and discussed articles focused on teaching for transfer (National Research Council 2005), rigor (Stein et al. 2000), student engagement (Schlechty Center on Engagement 2009), and assessment (e.g., Dufour 2004; Zmuda 2008). After each reading, teachers discussed ways to incorporate these ideas into their instruction. They also put these ideas to work as they designed tasks and assessments for their students.

Like the readings and discussions, most of the workshop activities related to participants' classrooms, and they provided opportunities for their own active learning (Darling-Hammond et al. 2009; Yoon et al. 2007). For example, student work from participants' classes was a primary focus in all workshops. Teachers generated student work by giving common assessment tasks, creating common rubrics, and bringing their students' work to the workshops. These tasks often had multiple solution paths and required students to solve nonroutine problems and explain their reasoning. The learning for teachers occurred as they worked together to analyze the student work and discuss student strengths and misconceptions. After identifying the difficulties, teachers worked in smaller groups to plan lessons that would take these strengths and weaknesses into account. Groups posted their lesson plans and other participants offered comments and questions about these plans, thereby allowing them to continue sharing ideas.

Through the process of gathering and examining student work, the college and high school instructors realized that student misconceptions were similar across the two levels. For example, on a functions transformation task that was based on the graph of a piecewise function labeled $y = f(x)$, students were asked to sketch the graphs of several new functions, including $y = f(x) - 4$ and $y = f(x + 2)$. Students at all levels sketched the parabolas corresponding to $y = x^2 - 4$ and $y = (x + 2)^2$, while others sketched the linear graphs of $y = x - 4$ and $y = x + 2$. In the first case, it is likely that the students who graphed parabolas had learned transformations in the context of quadratic functions and did not generalize their understanding to other functions; in the second case, students appeared to ignore the function notation completely. In both cases, students lacked common understanding of the meanings of function notation.

Professional Learning Communities Activities

The second group structure, the school PLCs, enhanced teachers' working relationships within their respective schools. Outside the workshops, PLC members expanded their collaboration by engaging in activities such as observing each others' classes and working together in a modified lesson study design to plan, teach, and evaluate "gourmet" lessons (and later, "gourmet" units). These lessons and units were a major activity for the PLCs, requiring great commitments of time. They also required teachers to be willing to take risks by sharing ideas and seeing those ideas in action. Many participants expressed a view that designing and teaching lessons together was challenging, but that it created a deeper sense of collaboration and trust among them. One teacher commented: "The collaboration piece is what's becoming important to me with this whole thing; it is just being able to collaborate within the building and then other people in the area and learn from each other."

The Randolph High School PLC's work on a gourmet unit illustrates the impact this work had on teachers and students. The three Randolph High School teachers and their postsecondary faculty member designed and taught a gourmet unit on conics. They combined and adapted ideas they had used before, and then brainstormed new ideas to create a two-week unit focused on the properties of conics that make them useful in real-world applications. Many of the student-centered lessons incorporated explorations on computers or with manipulatives. For example, a "thumbtack lab" allowed students to use thumbtacks and a string to investigate the properties of ellipses and the relationships between the properties.

The group also incorporated ideas from the workshop readings and discussions. One of these ideas was to include opportunities for students to reflect on what they had learned at the end of most lessons. These reflections were aimed at helping students take responsibility for their learning, a student attribute from the College Readiness Standards (TMP 2004). The final assessment included both a traditional paper-and-pencil exam and performance-based assessment projects. For final projects, students created parabolic cookers, elliptic pool tables, and art, and they presented these at a "Conics Carnival."

When the Randolph PLC presented this unit to the rest of the large group, one member of the PLC described his role in the classroom while teaching this unit as completely different from usual. Gary usually only lectured, having students take notes. He described the differences that occurred during this unit: his students worked together and asked each other questions rather than relying on him to explain, while his role consisted of facilitating students' work together and providing resources when necessary. He seemed genuinely glad to report that his students performed better on this unit exam than on any other exam during the school year. By planning and teaching a gourmet unit together, teachers in this group experienced different ways of teaching; they learned new instructional strategies with the support and insights of their colleagues, and they saw how their students benefitted from the experience.

Addressing Challenges

Several challenges arose during the project, and in each case, we adapted our planning to address them. Issues included dysfunctional groups, teachers' obstructive beliefs about their students, misunderstandings between high school and college participants, and participants' resistance to changing their instruction. In this section, we briefly describe each challenge and how we addressed them.

Dysfunctional Groups

In the first year of the project, there were just two high school members of the Carson High School PLC, Mike and Rene. Rene was not interested or committed to the project and so did not engage in ways that supported collaboration. For his part, Mike participated in and valued content-centered activities, but he disengaged from activities related to pedagogy. He described his teaching style as "stand and deliver" and explained that he felt alienated by his school district's push for more student-centered teaching. His students' preparation for college was very important to him, and he emphasized the need to prepare students to learn in the way their college instructors would teach, adding that all of his college mathematics instructors had lectured while students took notes. Project leaders noticed that the Carson PLC showed little commitment to the work, and that this group, along with others, lacked skills for collaborating effectively.

To address this, we introduced norms for collaborative discussions (Garmston and Wellman 1999). These norms included pausing after each speaker made a point, asking questions about the details of the idea and restating the idea before responding, maintaining a sense of inquiry and presumption that others had positive intentions, and paying attention to the way each person responded to the ideas. We revisited the norms at several meetings through discussions, role-playing, and sometimes humor, and we asked participants to reflect on how well their groups honored the norms. In most cases, this approach greatly improved the effectiveness of group work.

For the Carson team, the opportunity to work effectively came when Rene retired after the first year and two other teachers, Paul and John, joined the team. The new members embraced the norms and immediately showed commitment to project goals. In contrast to Mike's approach to teaching, Paul and John generally employed student-centered teaching practices. While these differences could have been an obstacle to group dynamics or further alienated Mike, they did not. In fact, the group members found their differences energizing. Mike described what he gained from their discussions:

> The opportunity to reflect and have disagreements. . . . Paul and I disagree about everything about teaching, and it sharpens both of us. I've really enjoyed that a lot. . . . Then, for me, the thing is to have confidence in Paul that we could disagree about things that were important to us but still value each other.

Because of the norms introduced in the workshops and practiced within the PLCs, Mike found a safe environment to present his views on teaching and learning. The group had found common ground in their concerns about how their teaching affected student learning and their willingness to improve their instruction. The norms helped them to form respectful ways of communicating and enabled them to explore ideas together in ways that allowed all members to learn. In spite of their differences, they depended on each other even outside of project activities. As Paul related: "[I like] being able to go to other people like Mike and say 'OK, this is what I'm trying to do, but I know it's not quite working, you know, [so] what's something you do that works for you?'"

By practicing the norms for collaboration, the Carson PLC laid a foundation that allowed them to be open to each other and to change. When they collaboratively planned and taught a "gourmet" lesson, they described it as particularly challenging but ultimately a tremendous learning experience. They attributed their learning to their differences and to their respect for each other. In contrast to his earlier resistance to changing his instruction, Mike initiated changes in his classroom based on the ideas addressed in the workshops. Classroom observations over three years showed that Mike had shifted from primarily lecturing to increased use of questioning in his classroom. In the most recent observation, the observer noted that Mike asked many more questions than in the earliest observations, that the questions focused on uncovering student thinking, and that Mike used students' answers to inform the direction of his instruction.

Tensions between High School and College Participants

In spite of our work on common understanding, goals, and norms, tensions sometimes arose that impeded group functioning. For example, a few high school teachers told us they felt intimidated by the presence of a university mathematician and by his aloofness from the group. They reported that he continued to work on mathematics even when the group shifted to other activities.

Our solution in situations like this was to explicitly address the issue one-on-one. In this case, one of the project leaders spoke privately with the mathematician, who expressed surprise that anyone would be intimidated by him, and emphasized that he valued getting to know high school teachers. His awareness of the teacher perceptions shifted his participation. In subsequent work with the group he interacted more with the teachers and attempted to keep teachers focused on mathematics (instead of the side conversations they were having about hobbies). Because of these changes, teachers later reported that he was approachable and an integral part of the project.

Teacher Beliefs about Students

Another issue that arose concerned teachers' beliefs about their students. Edison High School had the largest percent of low socioeconomic status (SES) students in the project, and the Edison PLC believed their students would necessarily perform lower on the common math tasks and would be less capable of persevering on challenging problems than students at the more affluent high schools.

What was most instrumental in shifting this belief was analyzing student work across sites, a design feature of our project. While listening to other groups share their analyses of student work on a problem-solving task, the Edison teachers noticed that their students had the same misconceptions as students at the higher-SES schools. During an interview, one of the team members recounted a discussion within their team:

> I think that it goes back to ... when you're at [Edison] specifically, you have like an inferiority complex. . . . Our scores aren't as good. . . . And then you see that everybody's having these same problems, and suddenly, I think the biggest thing that happens is that you can't say, "Well, our kids don't have money, well, our kids don't have this." Well, you know what, who cares? . . . All the kids are struggling with this thing.

The Edison PLC also noted that their most successful students were less likely to persevere on the problem-solving tasks than were their less-successful students. This realization provided insight into the teachers' instruction and student characteristics: they recognized that their regular classroom expectations did not require students to solve problems, so the students they deemed as successful had little experience in persisting when challenged. It also occurred to them that students they deemed less successful face many challenges in their lives, and through these challenges might have developed more persistence when tackling nonroutine problems. Thus, through discussing student work, these teachers began to change their beliefs about their students, generate new insights into their instruction, and become more committed to the goals of the project.

Changes to Instruction and Impacts on Student Learning

Early in the second year, through classroom observations, we realized that many teachers were not making changes to their instruction. Concerned about this, we identified four teachers who had made changes based on the project goals, and we asked them to describe their changes to the large group in a workshop. After the presentation, we asked each participant to commit to making a small change and to be prepared to share their experiences in the next workshop. Because we asked for only a small change, teachers seemed to accept this new request willingly.

Ideas for these changes often came from readings discussed in the workshops, and later from the little changes shared by other participants. Some teachers chose changes related to how they informed students about their learning—for example, by providing more specific feedback on student work. Other changes included ways to increase students' engagement in learning through teacher-student interactions and different assessment techniques. These changes entailed using small-group work to encourage communication about mathematics or having students present homework solutions or share their understandings of the "big ideas" of the lesson. Several teachers decided to work on their questioning strategies by asking higher-level questions and by responding to student questions with probing and prompting questions rather than answers.

A common change targeted the College Readiness Standards student attribute of perseverance. Lauren was one of the teachers who chose to work on improving her students' persistence in solving mathematics problems. In an online forum created for teachers to share their little changes, she described how she gave a mathematics task from the project to a couple of her freshman classes:

> I was convinced they could do it. I told them at the beginning of the class that I would not be providing any assistance on the task, that it would be challenging, but if they stuck to it, they could solve the problem. It was awesome. It took about twenty minutes for anyone to come up with even one solution, and at the end of class some students still had not found a single working solution, but those that found one (and a couple of students even found an infinite number of solutions) looked like they had just won a marathon. I remember being astonished how well many of the students did struggle, and in the end persevere. There were kids who quit before the time was up, but far more kept trying and were begging to hear solutions before they left. . . . The students felt vested in the problem—they wanted to know. I think that is part of students developing good struggling skills (is there such a thing?). . . . They need to want the result.

Like Lauren's small change, other teachers' small changes affected students' experiences in the classroom and also affected teachers' beliefs about students' abilities. When describing their small changes, teachers often also described resulting changes in students. For example, teachers who had students do more presentations of big ideas and homework solutions noticed that the students were developing stronger communication skills. Teachers who changed their forms of assessment noticed that their students became more articulate about what they did not understand, focused more on the objectives, and worked harder to meet them. Many teachers, like Lauren, realized their students could engage in more challenging work than the teachers had previously expected, and that students could enjoy the challenge.

Listening to other teachers' changes also inspired participants. At workshops after teachers were asked to initiate a small change, teachers shared their new practice with others through a "give one, get one" protocol. According to this protocol, each participant talked to ten other participants for a few minutes each, taking turns to describe their changes and asking clarifying questions. Participants recorded the small changes they heard and then rejoined their groups to share what they had learned. Very often while describing someone else's change, teachers would explain that they wanted to try this strategy in their own classes.

Susan, a college instructor, made changes that provided another example of student impact. Susan had historically used an approach of teacher lecture and demonstration. As her change, she asked students to compare their answers on homework with each other in class before she went over problems. When we interviewed her, she described both her change and the effect it had on students.

> Take that "I'm the all-knowing person" away from them and start having them talk. . . . So they have about ten, fifteen minutes every day at the beginning of the class to compare their answers. And so, they started talking to each other and comparing and things changed in the classroom.

Susan's classroom became much noisier, which made her uncomfortable at first, but, as she relates, "There's still the thing I like . . . They're getting excited about it; they have a lot of energy now in that class." The change in students' engagement further affected Susan. Several months after instituting this change in one class, she described her discomfort with the relative quiet of another class in which she had not instituted changes, and she was considering ways to increase students' active participation.

By early in the third year, participants began to see changes in their instruction as a primary goal of the project. This was reflected by the most common responses on workshop evaluations about how the project was affecting their work: "try new strategies," "becoming more intentional about pedagogy and do it consistently," "continual revision," "bridge gap between meetings and classroom practice," and "continued small change." Participants also saw how their small changes could improve students' mathematical processes and efforts to learn: "let students think/collaborate more," "more problem solving," and "model perseverance." Like Susan and Lauren, many of the teachers were making the connections between changing their teaching and changing the way students engaged in learning and in mathematical processes.

Project leaders also observed changes in students during classroom observations. In one of the case studies, one of us observed a high school teacher before and after he initiated his small change and noted, "There is a noticeable difference in the way his students behave in his classroom. They are taking more responsibility for their learning and for helping the others in their groups. He also has much more connection with each student, even though he is putting more of the responsibility on them." Another one of us overheard a student tell another student how much better she learned in groups after her teacher employed small groups for a review game.

For many teachers, a change in their students became the purpose of their little change. Karen, a high school teacher, incorporated several small changes over a two-year period, often incorporating ideas she learned in the workshop sharing protocol. In the third year, she described a change she adapted from one that had been shared. Her goal with this change was to get students to reflect on their own learning. She initiated it by explaining why she would be asking them to do something new. She told her students that to be ready to succeed in college mathematics, they would need to have the skills and dispositions to take responsibility for their learning. She supplied them with a list of learning objectives in clear student-friendly language, such as, "I can factor trinomials with a leading coefficient of 1" or "I understand and can explain the connections between a linear equation and its graph." Then on each homework assignment, Karen asked students to use a rating scale to assess their own understanding of each learning target. She emphasized that if they rated their understanding low, they needed to come to her for extra help outside of class. She was surprised at how seriously they took this task. Later, when she no longer required students to rate their understanding, she was surprised again that they continued to do it: "But where I thought that they would not do that, . . . they continued to, like, think about 'where am I with it?' I see that as a huge impact on them, and . . . and that was so small, that little piece." When pressed to explain the "huge impact," she said, "It's changing the culture of my classroom, that's the impact." Students were interested in taking control of their own learning.

Karen also described how the idea of small changes had affected her colleagues outside of the project:

> I think that that's one of the biggest impacts that has been brought back [from the project] to our schools and our team; it's the idea of little changes in doing small things within your own classroom. . . . I think it just really gives permission for people to try something without thinking that there has to be this huge planning of it or whatever. . . . I approached that with my department ... and I could just see this huge kind of relief come out of them.

Karen and others had been empowered to change their instruction and had seen its effect on their students, so they shared their enthusiasm and a plan for implementing change.

In summary, we were challenged throughout this project to create and maintain a community of faculty and teachers from high schools and colleges who would take ownership of a problem, develop common understandings of it, and generate solutions. While the norms laid the foundation for ways of collaborating, the activities of individual teachers, the PLCs, and the large group helped deepen the conversations and support change in both teacher learning and student learning. Although we structured the PD activities with specific overarching goals in mind, the design honored teachers' ideas, encouraged them to be open to each other's ideas, and empowered them to try new ways to improve student learning.

......

This project is funded by a grant authorized by the Elementary and Secondary Education Act and administered by the U.S. Department of Education and Washington State Higher Education Coordinating Board and by additional funding from the Washington State Transition Mathematics Project.

REFERENCES

Attewell, Paul, David E. Lavin, Thurston Domina, and Tania Levey. "New Evidence on College Remediation." *The Journal of Higher Education* 77, no. 5 (September/October 2006): 886–924.

Common Core State Standards Initiative. *Common Core State Standards for Mathematics. Common Core State Standards (College- and Career-Readiness Standards and K–12 Standards in English Language Arts and Math).* Washington, D.C.: National Governors Association Center for Best Practices and the Council of Chief State School Officers, 2010. http://www.corestandards.org.

Darling-Hammond, Linda, Ruth Chung Wei, Alethea Andree, Nikole Richardson, and Stelios Orphanos. *Professional Learning in the Learning Profession: A Status Report on Teacher Development in the United States and Abroad.* Dallas, Tex.: National Staff Development Council, 2009.

DuFour, Richard. "What Is a 'Professional Learning Community'?" *Educational Leadership* 61, no. 8 (May 2004): 6–11.

DuFour, Richard, and Robert Eaker. *Professional Learning Communities at Work: Best Practices for Enhancing Student Achievement.* Bloomington, Ind.: Solution Tree, 1998.

Frost, Janet Hart, and Jacqueline Coomes. "Meeting Teachers Where They Are: Findings from a High-Density, Low-Pressure Mathematics Professional Development Project." Paper presented at American Educational Research Association Annual Meeting, Denver, Colo., May 2009.

Garmston, Robert J., and Bruce M. Wellman. *The Adaptive School: A Sourcebook for Developing Collaborative Groups.* Norwood, Mass: Christopher-Gordon Publishers, 1999.

Gordon, Sheldon P. "What's Wrong with College Algebra?" *PRIMUS* 18, no. 6 (2008): 516–41.

National Research Council (U.S.), M. Suzanne Donovan, and John Bransford. *How Students Learn: Mathematics in the Classroom.* Washington, D.C.: National Academies Press, 2005.

Schlechty Center. "Schlechty Center on Engagement." 2009. Retrieved from http://www.schlechtycenter.org.

Stein, Mary Kay, Margaret Schwan Smith, Marjorie A. Henningsen, and Edward A. Silver. *Implementing Standards-Based Mathematics Instruction: A Casebook for Professional Development.* New York: Teachers College Press, 2000.

Transition Math Project. *College Readiness Standards in Math.* 2004. http://www.transitionmathproject.org/standards/doc/2010-crs-16feb10-revisions.pdf.

Yoon, Kwang Suk, Teresa Duncan, Silvia Wen-Yu Lee, Beth Scarloss, and Kathy L. Shapley. *Reviewing the Evidence on How Teacher Professional Development Affects Student Achievement* (Issues & Answers Report, REL 2007-No. 033). Washington, D.C.: U.S. Department of Education, Institute of Educational Sciences, National Center for Education Evaluation and Regional Assistance, Regional Educational Laboratory Southwest. http://ies.ed.gov/ncee/edlabs/regions/southwest/pdf/REL_2007033.pdf.

Zmuda, Allison. "Springing into Active Learning." *Educational Leadership* 66, no. 3 (November 2008): 38–42.

CHAPTER 17

Voices of Mathematicians and Mathematics Teacher Educators Co-Teaching Prospective Secondary Teachers

Denisse R. Thompson
Catherine Beneteau
Gladis Kersaint
Sarah K. Bleiler

BOTH MATHEMATICIANS AND MATHEMATICS TEACHER EDUCATORS (MTEs) have responsibility for preparing mathematics teachers. In many institutions, mathematics content courses are taught by mathematicians, and mathematics pedagogy courses by MTEs. Often in separate departments or colleges, these two groups may live in different worlds with different cultural norms. The Conference Board of the Mathematical Sciences (CBMS 2001) notes: "There is considerable distrust between mathematics faculty and mathematics education faculty both within institutions and through public exchange. Conscious efforts ... are needed to foster cooperation, along with mutual understanding and respect" (p. 9).

Public discourse about mathematics teacher preparation is often based on content knowledge. Yet as Ball (2003) acknowledges, "Increasing the quantity of teachers' mathematics coursework will only improve the quality of mathematics teaching if teachers learn mathematics in ways that make a difference for the skill with which they are able to do their work." The effort to engage teachers in learning mathematics in ways that resonate with their future teaching expectations suggests a need for collaboration between mathematicians and MTEs as recommended by the CBMS:

> Most good school mathematics instruction involves a combination of mathematical knowledge and pedagogy. . . . Mathematics educators can provide valuable insights and information about what takes place in school classrooms, including common mathematical misunderstandings of practicing teachers. . . . [M]athematics faculty can help mathematics education faculty by keeping them informed of mathematical developments which have an impact on school mathematics. (p. 9)

Collaboration can take many forms, from sharing of ideas and philosophies in occasional discussions (Ball et al. 2005) to developing curriculum materials for professional development with teachers (Kersaint and Berger, this volume) to co-teaching of courses (Grassl and Mingus 2007). In this chapter, we share our experiences in creating a learning community from a collaboration among a mathematician (Catherine), two MTEs (Denisse and Gladis), and a mathematics education doctoral student (Sarah) who collaborated on the development and delivery of a geometry course required of all secondary preservice teachers (PSTs).

We start by describing the context for the collaboration and some practical logistics, including our preparation for class and the experiences we designed for PSTs. Then, we each offer our perspective on the learning community that emerged among the four of us. We also share insights from PSTs collected during a focus group discussion about how they viewed the roles of the two instructors and how their experiences in our class compared to their experiences in other mathematics courses. We conclude by looking across the perspectives of the four collaborators for common views and issues related to such endeavors.

Context for the Collaboration

The University of South Florida is a research university with over 47,000 students. Although the mathematicians and MTEs coordinate class schedules to avoid conflicts for PSTs, more in-depth collaborations have only begun within the last five years. Gladis has had several grants in which mathematicians, including Catherine, have engaged to develop and deliver professional development for K–12 teachers. This work laid the seeds for the collaboration discussed here.

Our university is one of four universities that were participating in a National Science Foundation–funded project based at the University of Arizona. This project was designed to study the nature of collaboration that occurs when a mathematician and a MTE co-teach a mathematics content course and a mathematics pedagogy course. (The project was NSF DR K-12 No. 0821996, Knowledge for Teaching Secondary School [KnoTSS], with Principal Investigator Rebecca McGraw.) Although Gladis secured our participation, she was not able to serve as co-instructor, so Denisse fulfilled this role. Hence, an opportunity arose for a unique collaboration of four individuals, with Catherine and Denisse as co-instructors for two courses, and with Gladis and Sarah participating in the planning and observing most classes during the fall 2009 semester.

Neither Catherine nor Denisse had ever previously taught this college-level geometry course, nor had they previously collaborated in any way. The course met for seventy-five minutes twice each week.

Overall Goals of the Course and Collaboration

When we first met to design a syllabus, we established some objectives that would permeate the course. Specifically, PSTs should (a) learn mathematics using inquiry-based approaches as recommended by the mathematics education community (e.g., Martin [2007]; NCTM [2000]); (b) reason about and make sense of mathematics for themselves, often within a structure of collaborative groups; and (c) write mathematical proofs, and use the language of mathematics appropriately.

Throughout the semester, the four of us met at least once per week for two hours to prepare for upcoming classes. We typically began by sharing insights from the previous week's classes, specifically sharing observations about instruction, concepts that seemed to cause difficulty, or particularly noteworthy comments from PSTs. We then discussed the content for the week and created PowerPoint outlines in which we indicated activities that were to occur, theorems to be stated or proved, and teaching notes for ourselves regarding important points. The planning meetings were an opportunity to brainstorm inquiry-based activities for engaging PSTs with the content and to raise concerns or questions about the content itself. In every class, we made an effort to incorporate some kind of group work.

To ensure that both Denisse and Catherine truly co-taught the course and shared significant instructional roles in each class period, we designated in our notes who would lead each segment of the lecture or class activity. Thus, we created an expectation that Denisse and Catherine both had an important contribution to make during every class.

A Look into the Class

First day. Setting the tone for the semester on the first day of class was crucial. We wanted to establish both for ourselves and for the PSTs that mathematics as well as education issues would be present, and that a social environment would be created in which the PSTs would discuss the content and work with each other. So, we began this class by having the PSTs put themselves in order according to their birthdays without speaking. Such an activity strongly hinted that this class might be different from what they normally experienced in mathematics! We had the PSTs discuss why we had engaged in such an exercise, and they raised issues of communication, group work, and their own expectations for a mathematics class. The activity provided a shared experience for all to serve as a foundation for what would occur throughout the semester.

A typical class day. For each class, PSTs were expected to read some portion of their text prior to class, often focusing on key definitions related to geometry. Weekly, we gave a brief quiz to support this reading and also collected homework. It was typical to begin class with feedback from a previous class, such as topics from a previous lecture that seemed unclear to PSTs or common mistakes in the homework or on the quizzes. We then introduced the day's work or activity, such as examining or discovering definitions or proofs of theorems.

Sample strategies. Throughout the semester, we experimented with various activities to engage PSTs during group collaboration. Because geometry is highly visual, we often used patty paper to illustrate concepts or to support developing proofs. On one occasion, PSTs had read prior to class about important lines in a triangle (i.e., median, altitude, and angle bisector). Given a triangle, they then used patty paper to construct a median of a triangle. They folded the patty paper so that two endpoints of one of the sides of the triangle matched and made a crease; then the point at the crease represented the midpoint of that side. As PSTs drew the segment from the opposite vertex to this midpoint, their actions reinforced the distinction between "midpoint" and "median" in a concrete manner without that distinction needing to come from either instructor. On another occasion, we used patty paper to model the side-angle-side congruence test; by physically moving triangles and lining up sides and angles, PSTs had a hands-on experience of the meaning of this important congruence test.

Another particularly successful strategy that emphasized the importance of communication was to have PSTs anonymously critique some written work of their peers. Specifically, we selected sample responses to a quiz or a homework problem and prepared these on a handout. PSTs worked in groups to evaluate the responses according to the following guidelines: (1) Were the responses correct, clear, and complete? And (2) If the response was incorrect, could it be suitably modified? After PSTs discussed these sample responses within their groups, we had a whole-class discussion of their critiques. This exercise seemed to help PSTs understand our expectations for clear written work without those criticisms coming directly from us. The PSTs were typically thorough in their critiques, so the instructors rarely had to make additional comments but simply had to guide the discussion for clarity. At the end of the exercise, PSTs had coherent, concise, and correct answers to the given questions and ownership over the material in a way that might not have occurred had they simply received a corrected paper.

Grading and evaluation. Denisse and Catherine both fully participated in grading the PSTs' work. Although grading of assignments alternated between instructors, they discussed the graded papers and made any adjustments before returning assignments. This shared responsibility gave each instructor ownership of the course and was key to the collaboration being significant. PSTs generally could not tell who graded their assignments.

Sharing Our Perspectives on the Collaboration

In the following four sections, we share our individual perspectives on the collaboration. We start with the two co-instructors (Catherine, the instructor of record, and Denisse), followed by the two mathematics educators who observed the class (Gladis and Sarah).

Catherine: A Mathematician Co-teaching the Course

I had never team-taught a course before; sharing the privacy and intellectual domain of the classroom was initially difficult. Having observers (Gladis and Sarah) made me particularly self-conscious, and I had to adjust to receiving constructive criticism. Because I had not taught geometry before, I had few preconceived notions about the course itself and was willing to keep an open mind about topics to cover or strategies to try. After a few weeks, I began to enjoy the natural cycle of feedback and discussion about what went well and what didn't. One day when Denisse couldn't make it to class and the others were late, I was disappointed to be on my own; I had become accustomed to sharing the classroom and bouncing ideas off her during class.

One of the things I most enjoyed about the collaboration was the preparation for class. We each reviewed the mathematics in the sections we planned to cover before our meeting, shared ideas about what concepts were important, and brainstormed about how they should be taught and the activities we might incorporate. We spent time working on challenging geometry problems. Because we were using a traditional mathematics text, we struggled to identify ways of teaching the material that would support an inquiry approach. I found the mathematical and pedagogical challenges of the preparation intellectually stimulating.

Before I began this collaboration, I considered myself a "good" teacher and someone who connects to students and recognizes what they do not understand. I had previously experimented with various strategies (e.g., group work, technology), but I feel that there was something fundamentally different about the initial setup of this geometry class. Although the strategies we employed were not that different from those I had implemented on occasion in other classes, they were incorporated as a fundamental piece of the geometry course design.

This collaboration has given me numerous concrete ideas about how to deliver instruction with a student-centered approach, such as having students use patty paper in geometric constructions, or spaghetti to discover the triangle inequality. One specific technique I found useful was to create a handout of students' quiz responses for them to critique. My role shifted to discussion leader, and criticisms came from students, who were able to judge, with a little guidance, the coherence and correctness of their peers' responses. Through such activities, students became more critical during the semester, better able to evaluate peers' responses, and more adept at writing their own responses in future assignments. In addition, Denisse and I gained access to what the students really understood. It became clear that the students were having great difficulty in identifying the hypothesis and conclusion of a given mathematical statement, and we needed to spend time helping them understand references to any pronouns in the statement of a theorem before we had any hope of teaching students to prove it.

I also became familiar with what students themselves will face in the classroom in terms of content, and more importantly, I gained a sense from my collaborators about what is emphasized in the high school curriculum, such as the use of congruence tests for triangles. This is difficult for a mathematician to know if working alone.

Throughout the semester, I became more aware of how my own imprecise use of notation or language might affect student learning. For example, I was not careful to write relationships for similar triangles so the vertices corresponded to the angles that are congruent. Although I would never mark a response wrong if the correspondences were not written in the proper order, I had not spent much time thinking about how such notation might confuse students.

Finally, the time commitment was tremendous. I wouldn't have been able to participate in this collaboration before I applied for tenure, because it took time away from more traditional mathematics research activity. Although my chair has been supportive, in terms of "research deliverables" that carry weight in a mathematics department, this project might not be considered valuable. On a personal level, however, the project has been stimulating and has changed my view of myself as a teacher and what my classroom could be. I believe this collaboration might be a starting point for interesting discussions in my own department.

Denisse: A Mathematics Teacher Educator Co-teaching the Course

Geometry has never been my favorite area, even though I taught high school geometry one year. So, being responsible for teaching while being observed was a bit unsettling at the beginning. I felt like judgments might be made about my mathematical knowledge. Even though I had previously team-taught a geometry course for in-service teachers with Gladis, there was still anxiety at the beginning, particularly when the observers saw something that they thought could be improved.

I was concerned about entering into this collaboration because I felt like the odd person out. Gladis and Catherine already had a working relationship; Catherine and Sarah also had a relationship because Catherine had been Sarah's master's thesis advisor. I mention this initial unease because developing trust and mutual respect are essential to an effective collaboration and this develops over time. It builds gradually as each team member shares during planning and potential instructional ideas are discussed, validated, modified, and respected.

I completed my mathematics coursework well before the implementation of the standards movement from NCTM. Even though I try to teach mathematics pedagogy courses with discussions, cooperative group work, and problem solving, I have not had an opportunity to implement those practices into a mathematics content class. I was faced with the practical reality that many of the recommended pedagogical approaches are challenging to implement, especially when so much content needs to be addressed. Deciding to give students more time to discuss a problem in their groups, even though it means that the planned lesson is not finished, is difficult. Deciding that completing fewer content chapters is okay is hard, even though I believe that students have a better mastery of the content that we covered. The goal of having PSTs become careful about their language when writing mathematics meant we needed time to discuss and critique samples of their writing if they were going to improve. The time spent on this activity had to be taken from elsewhere. It is easy to talk about "less is more" from an abstract perspective; it is harder to put that perspective into practice when you are responsible for teachers' mathematics content knowledge.

I am pleased that our collaboration resulted in a classroom in which students regularly engaged in mathematical conversation and in which they experienced learning content using practices I typically discuss in methods classes. Catherine brought good mathematical insight into our discussions and helped ensure that the cognitive demand of the tasks remained high. Even when the PSTs were struggling with content, Catherine was not willing to lessen her expectations but instead wanted to work together to implement strategies (e.g., the critique of quiz responses) that could help the PSTs overcome those struggles on their own. While wanting to ensure that the essential content was addressed, she was more concerned with PSTs' learning than with rushing through content for the sake of content coverage. Because of past discussions with students, this was not what I had expected from the mathematicians at my university.

I feel I was an equal partner in teaching the geometry course, but the time commitment was huge. I was not technically listed as an instructor of record. Because we were participating in the grant, I had a course release during the year. But without such a release, could I afford to spend the time that such collaboration requires? In mathematics education, we teach the same courses on a regular basis, so time commitments for intense reflection and course modification pay off. In contrast, in the mathematics department, courses circulate among various faculty members. This means I could invest the time to suggest improvements in one of the content courses taken by our PSTs but it might be for naught when someone else teaches the course in the next semester. Regardless of the outcome, I learned that it is important for MTEs to take an active role in enhancing mathematics content courses, particularly those taken by prospective teachers.

Gladis: A Mathematics Teacher Educator Observing the Course

Unlike Catherine and Denisse, I spent several years teaching high school geometry using many of the strategies recommended for instruction by the mathematics education community, including the use of technology. So I was looking forward to my role as a co-instructor and to incorporating those strategies in teaching a university content course. Although I couldn't be a co-instructor, I thought it would still be possible to incorporate many of those approaches through my participation in the planning. However, I found this to be a challenge.

Because I was familiar with all the players, I assumed trust would exist from the onset. Denisse and I had collaborated in various capacities, including co-teaching a course, co-authoring manuscripts, and collaborating on programmatic issues. Catherine had been involved in a prior grant effort to develop and deliver content-specific professional development which resulted in my observing her teaching. Sarah was a doctoral student in one of my courses. However, I failed to anticipate the need for others to cultivate relationships. This was clearly an oversight, given that Catherine and Denisse met only to engage in this effort. In hindsight, it makes sense that each approached this effort with some trepidation.

During initial planning meetings, I shared my observations from class sessions to facilitate planning for future sessions. As an observer, I was able to focus on instructional issues, instructor-student and student-student interactions, and the nature of the learning environment. I was surprised that both Denisse and Catherine found this intimidating and initially viewed observational comments as judgmental. Denisse and Catherine had no prior working relationship and were grappling with the nature and content of the course (e.g., how it was addressed in the selected text, and what content to emphasize). Having an observer comment on every aspect of the classroom was jarring when they were trying to figure out how best to work with each other. Consequently, I had to adjust my role to allow the instructors to make sense of the course on their own terms, and I became selective in observations and strategies I shared. In earlier stages, I suggested instructional approaches that were familiar to me. However, on occasion their use in the classroom revealed a lack of shared meaning among the group that was not apparent during the planning meeting. Although I knew what I would do, say, ask, and emphasize if I were teaching, that information did not transfer as I explained it to the others. At times, Sarah and I discussed whether what was observed represented what had been shared during planning. We both acknowledged that these perceived differences were differences in interpretations rather than incorrect uses of any strategy. Viewing these differences highlighted the need for shared meaning among the collaborators during planning. This required that I step back and ask questions to determine what individuals were making of the discussions, such as "What is the goal?" and "How do you see this playing out in the class?" Responses to these questions revealed differences in interpretations and provided opportunities to clarify understandings.

I am pleased with this collaboration. The classroom environment, students' feedback, and collaborator feedback have been positive. This collaboration has reinforced previous work in engaging mathematicians meaningfully in the work of teacher education. There is greater appreciation of the role both groups play. These initial efforts will broaden the discussion among MTEs and mathematicians and build the foundation for other efforts to improve the mathematical education of teachers.

Sarah: A Mathematics Education Doctoral Student Observing the Course

I have studied in both the mathematics and the education departments at this university. In universities like ours, PSTs often experience a disjointed education. Mathematicians are responsible for PSTs' learning of mathematical content, and MTEs are responsible for PSTs' learning of mathematical pedagogy. However, it is not often made clear to PSTs how the content of their university mathematics courses, seemingly disparate from the high school mathematics curriculum, relates to their future classrooms. Within this course, Denisse was able to point out issues related to the content in high school geometry classes that would be important for PSTs to know. For example, during a discussion about quadrilaterals, Denisse noted that the definition of *trapezoid* may differ depending on the text that teachers use, an important consideration given the pressures for accountability during high-stakes testing in which definitions on tests may be inconsistent with definitions in curricular materials.

The PSTs were not only able to make connections between the content of high school and university-level mathematics, but were also able to experience inquiry-based instruction within an actual mathematics class, a type of instruction many PSTs have rarely seen put into practice within mathematics courses. As a student new to the field of education myself, and accustomed to the traditional teacher-centered, lecture-style instruction within the mathematics department, I often feel overwhelmed by the recommendations for inquiry-based teaching and learning. What exactly does it mean to teach in an inquiry-based fashion? Had I ever seen a teacher who taught that way? What would it look like? I can't help but believe PSTs ask themselves the same questions. Within this course, PSTs were able to experience firsthand an inquiry-based classroom. For example, after working on a problem collaboratively in small groups, Catherine and Denisse would pull the PSTs back together for a whole-class discussion in which the PSTs' contributions, as opposed to the instructors', would determine the flow and direction of the discussion. PSTs listened closely to each other, focused on the accuracy and precision of mathematical language used within the classroom, and questioned each other when something was unclear or seemingly incorrect.

I believe this type of classroom environment was successfully cultivated as a direct result of our collaboration. Our differing levels of experience as teachers, mathematicians, and teacher educators contributed to the variety of perspectives through which we viewed the course. As MTEs, Gladis and Denisse identified activities that would help the PSTs discover mathematical relationships on their own and develop precision in their mathematical language. As a mathematician, Catherine recognized connections among foundational aspects of the subject that helped lead the PSTs to derive formulas instead of simply memorizing them (e.g., the law of cosines). As someone new to education, I was eager to learn about pedagogical tools and activities teachers could use in their own classrooms, and therefore researched and proposed several such activities (e.g., breaking up spaghetti into three pieces to formulate a conjecture about the triangle inequality). While Catherine and Denisse were teaching, Gladis and I made observations and discussed the unfolding events. We viewed the instruction from the perspective of the instructors and also focused our attention on the reactions of the PSTs to the instruction. During our weekly meetings, the four of us discussed class sessions from our different perspectives and used these reflections to design the next class accordingly.

I gleaned several personal lessons from participating in this collaboration. Through interaction with and observation of secondary PSTs, I made valuable connections between research and practice. Because my current research interests focus on the teaching and learning of mathematical proof, being in the classroom with PSTs and reflecting on their learning of proof in geometry helped bring readings from research to life.

Prior to this collaboration, I would have resisted having someone observe my classroom. But now, I welcome the opportunity to reflect on my teaching practices with others because I see the value that such discussion and reflection brought to Catherine's and Denisse's instruction.

Learning from Our Students—Preservice Secondary Teachers (PSTs)

In such collaborations, one might expect the MTEs to contribute most of the pedagogical strategies in the course while the mathematician focuses on aspects of the content. However, at least for us, this was not the case. Catherine had valuable insights about student learning and how to enhance that learning within mathematics classes, and Denisse regularly contributed insights related to content. One of our PSTs noticed this dynamic:

> I think Dr. Beneteau's extensive teaching in the math department gives some pretty good insight into how her students learn. It might be different material . . . calculus and analysis, but she still knows how students learn, what questions they have, how to help them during office hours, during discussions, how to teach them. She teaches well.

One student voiced concern over the style of instruction in the course, illustrating a preference toward the familiar lecture-style model of instruction over the inquiry model we used. Others (e.g., Rodriguez and Kitchen [2005]) have documented PSTs' apprehension/ resistance toward inquiry-style learning environments, which is typically linked to years of enculturation within traditional lecture-style mathematics courses.

> That's [the style of instruction] one of the few things that I did not like about the geometry course. I like math lectures. I don't like group work, I don't work well with others, I like to show up, do my stuff, learn my stuff, be tested on my stuff, and collect my A.

Other students, however, appreciated the opportunity to learn geometry through inquiry-based methods and group work.

> I liked geometry when we did group work. I'm sorry, I'm the kind of person that needs to be engaged . . . Like math lectures, I fall asleep. I don't pass the class. I don't do the homework because I don't care about the class. The geometry homework was a lot more engaging because I enjoyed that class a lot and I feel like it was because of the teaching style. I liked it because it was different, because it was group work and stuff.

Personal insights gained from the inquiry perspective were also evident in one particularly poignant instance. During a class focusing on the distance formula in the plane, PSTs were guided to determine the distance along horizontal and then vertical segments. They then considered what they knew about the Pythagorean theorem to determine the distance along an oblique segment. After the typical distance formula was determined by the PSTs on their own in their group, one PST doing quite well in the class commented, "I feel so dumb." In discussing her comment with the class, she indicated that she had never made the connection between the distance formula and the Pythagorean theorem; she had simply memorized the distance formula and often got it wrong. Her comments and the subsequent discussion provided a unique opportunity to discuss the value of understanding over rote memorization; when one knows how a formula is generated, he or she can quickly reconstruct it when needed.

To many of the PSTs, the lines between co-teaching and inquiry teaching were blurred because the incorporation of a mathematics educator within the mathematics class provided opportunities for them to see and learn the subject in new ways. One PST compared Denisse's presence in the geometry course to previous experiences she had with her in education courses:

> She [Dr. Thompson] did some of the things she does also in the methods class, that in a straight math class I've never been exposed to, group work in a college math class . . . the way that she analyzed our proofs on the board. Having the students write them and then going through them line by line to see what's right, what's wrong, how to improve the proof. I did learn a lot about geometric proof writing in that course, and I think Dr. Thompson had a lot to do with that.

Conclusion and Discussion

To improve the mathematics preparation of future secondary teachers, mathematicians and mathematics teacher educators must work together. All four of us believe the collaboration and the learning community that we developed among ourselves has enhanced our perspective of the role each of us plays in preparing future teachers. As we reflect on this co-teaching collaboration, common perspectives permeate our individual narratives.

We needed time to develop respect and trust and to build a way of working together. Catherine needed time to recognize that Denisse was both mathematically competent and interested in engaging in the mathematics. Denisse needed time to realize that Catherine was willing to try different pedagogical strategies, even if she was not sure they would work.

The collaboration supported the development of an inquiry classroom (e.g., a constructivist approach) within the learning of a content course. Catherine learned what constructivism was, Denisse was able to implement methods within a content course that she discusses in pedagogy courses, and Sarah witnessed how research about constructivism can be put into practice. Catherine had a chance to realize the benefits of a classroom that engages students in doing mathematics; Denisse and Gladis experienced some of the challenges in implementing recommended instructional approaches.

In order for our collaboration to work smoothly and seamlessly, we had to develop shared meanings and a common language. For instance, acronyms and references like NCTM or the *Standards* are used regularly by mathematics educators, but they are not necessarily understood by mathematicians.

Because neither Catherine nor Denisse had previously taught this geometry course, neither came to the co-teaching with preconceived notions of what the course should look like. They were creating a course together. As indicated in the narratives, Gladis had a long history of teaching geometry and initially had a vision of what should take place instructionally at certain points; as noted, this created some initial discomfort which had to be worked through.

We each had to give up some of our personal space within the classroom as part of the collaboration. Typically, instructors are the ones in control. Because of the collaboration, we each had to share the pedestal and the relationship that instructors build with their students.

The time commitment was immense, much more than either Denisse or Catherine expected. They both engaged in all stages of the collaboration, including sharing teaching, planning, and grading. Through grading, each instructor developed relationships with the PSTs, learned about their strengths and misconceptions, and was able to determine further instructional practices to use. The time commitment resulted in more reflection about the course than often occurs when one teaches alone, and this reflection had to occur during time when all parties were available. Although we believe in the collaboration, trying to make it happen without additional time (as provided through the grant) is difficult. Such efforts are not incentivized by our university's reward structures, and they take time from other areas (i.e., research) that are rewarded.

We all believe we grew professionally. Denisse and Gladis gained a better appreciation for the concerns that Catherine and other mathematicians have about PSTs' mathematical knowledge. The openness we developed in talking about the mathematical background of our PSTs and their needs provides a foundation on which to build collaborations to improve other courses taken primarily by PSTs. As documented, the PSTs generally had positive perceptions of the course developed through this collaborative endeavor.

Although this paper discusses our learning from a single (fall) semester of collaboration, we continued the collaboration during the spring semester in a mathematics pedagogy course. Issues of building trust and working through initial tension did not arise the second semester, and the collaboration was smoother. Thus, continued collaboration has the potential to enable more instructional innovations to be incorporated into teacher preparation programs in both content and pedagogy. The challenge is to develop support to institutionalize such endeavors.

REFERENCES

Ball, Deborah L. "Mathematics in the 21st Century: What Mathematical Knowledge Is Needed for Teaching Mathematics?" Paper presented at the Secretary's Summit on Mathematics, U.S. Department of Education, Washington, D.C., February 2003. http://www2.ed.gov/rschstat/research/progs/mathscience/ball.html.

Ball, Deborah L., Joan Ferrini-Mundy, Jeremy Kilpatrick, R. James Milgram, Wilfried Schmid, and Richard Schaar. "Reaching for Common Ground in K–12 Mathematics Education." *Notices of the American Mathematical Society* 52, no. 9 (June 2005): 1055–58.

Conference Board of the Mathematical Sciences. *The Mathematical Education of Teachers: Part I.* Washington, D.C.: Mathematical Association of America, 2001.

Grassl, Richard, and Tabitha Mingus. "Team Teaching and Cooperative Groups in Abstract Algebra: Nurturing a New Generation of Confident Mathematics Teachers." *International Journal of Mathematical Education in Science and Technology* 38, no. 5 (July 2007): 581–97.

Kersaint, Gladis, and Sandy Berger. "Negotiating Cultural Meanings: A Large-Scale Collaboration among Mathematicians, Mathematics Teacher Educators, and Teachers." In *Professional Collaborations in Mathematics Teaching and Learning*, 2012 Yearbook of the National Council of Teachers of Mathematics (NCTM), edited by Jennifer Bay-Williams and William Speer, pp. 259–70. Reston, Va.: NCTM, 2012.

Martin, Tami S. (ed.). *Mathematics Teaching Today: Improving Practice, Improving Student Learning.* 2nd ed. Reston, Va.: National Council of Teachers of Mathematics, 2007.

National Council of Teachers of Mathematics (NCTM). *Principles and Standards for School Mathematics.* Reston, Va.: NCTM, 2000.

Rodriguez, Alberto J., and Richard S. Kitchen (eds.). *Preparing Mathematics and Science Teachers for Diverse Classrooms: Promising Strategies for Transformative Pedagogy.* Mahwah, N.J.: Lawrence Erlbaum Associates, 2005.

Part VI

Going to Scale

How can collaborations support large numbers of students, teachers, and schools?

Formal and informal professional collaborations happen on a regular basis within school and work sites. Often school systems look for ways in which the "good work" that comes from these professional collaborations can support larger numbers of students, teachers, or schools—larger numbers than the numbers associated with the initial collaborations.

The three chapters in this section describe collaborations that have gone beyond the initial collaboration. We learn about how an idea that emerged from lesson study groups in two school districts spreads to twenty school districts and how an idea around problem-based mathematics becomes a statewide initiative.

As you read the chapters in this section we invite you to consider:

- What are the key factors and ideas that need to be in place in order for professional collaborations to support large numbers of students, teachers, or schools?

- What are the key factors and ideas that need to be in place in order to help support informal or formal professional collaborations to be broadened to more schools or teachers?

CHAPTER 18

Learning to Use Student Thinking

Development and Spread of "Re-engagement" Strategies Through Lesson Study

Catherine Lewis
Rebecca Perry
Shelley Friedkin
Linda Fisher
Jacob Disston
David Foster

MAKING STUDENTS' MATHEMATICAL THINKING VISIBLE during classroom instruction is a central challenge of mathematics teaching (Lesh et al. 2000; Boyd et al. 2003; Fisler and Firestone 2006; Ball and Bass 2000). This chapter chronicles how educators in a California lesson study network refined and spread "re-engagement," a teaching strategy designed to make student thinking visible and usable in classroom practice by presenting and discussing students' solutions to tasks. The chapter also examines the characteristics of the regional professional learning community that allowed knowledge about "re-engagement" to develop and spread.

Background on the Silicon Valley Mathematics Initiative and Lesson Study

This case focuses on the San Francisco–Silicon Valley region of California, where the Silicon Valley Mathematics Initiative (SVMI) supported lesson study as one strand of multifaceted work that also included an assessment collaborative and regional network of coaches. In lesson study, a small team of teachers studies the curriculum; plans a "research lesson" that is taught by one team member while the others observe students and collect data; and then discusses the collected data, drawing out the implications for teaching and learning (Wang-Iverson and Yoshida 2005; Lewis 2002; Gorman, Mark, and Nikula 2010). Evidence suggests that lesson study can have a positive effect on teacher and student learning (Perry and Lewis 2010; Lewis, Perry, and Hurd 2009; Fernandez and Yoshida 2004; Lewis et al. 2006; Foster and Poppers 2009). However, little has been written to date about lesson study as a means to support the spread of knowledge across a region or nation. This is a key aspect of lesson study in Japan, where a widely shared knowledge base for teaching has been developed and spread through lesson study (Lewis and Tsuchida 1997).

Overview of the Case

To understand the contours of the case, it is useful to compare lesson study in 2001 and in 2009 in the greater San Francisco-Silicon Valley region. In 2001, a small number of teachers in two districts within this region—Berkeley Unified School District (BUSD) and San Mateo-Foster City School District (SMFCSD)—had independently initiated lesson study groups, but they had not yet collaborated across district lines to observe live instruction. An inspection of research lesson plans from 2001–02 indicates that none of the lesson plans prepared by BUSD and SMFC teachers included presentation and discussion of student work as a way to introduce and consider mathematical ideas. Rather, the lessons were structured around teacher-developed or textbook activities intended to lay the groundwork for "correct" student thinking.

By 2009, teachers in more than twenty districts in the region had initiated lesson study, and many hundreds of teachers had observed and discussed lessons taught by educators from other districts. During every year from 2004–05 through the present, lesson study teams from fourteen to twenty-one districts chose to participate in a regional lesson study network that included public research lessons and exchange lessons in which lesson study teams partnered to observe each other's research lessons. By 2009, research lesson plans routinely anticipated student thinking, described the particular student work to be presented during the lesson, and noted the mathematical points to be drawn out. Teachers in at least seven districts were systematically testing and refining methods to present and discuss student work.

While recording research lessons in SMFCSD and BUSD, we noticed that teachers sometimes credited colleagues in other districts for elements of lesson design. In order to understand how ideas about mathematics teaching had spread across districts, we selected one element, "re-engagement," mentioned in research lessons in SMFCSD and BUSD, and we traced it backwards from 2009 to 2002. We discovered that it had spread across many districts of the San Francisco–Silicon Valley region; we limit the current account to seven lesson study groups in four different parts of the region, in order to use the data (lesson plans and meeting and interview transcripts) available to us. We focus first on the nature of re-engagement, and how teachers developed, refined, and adapted this strategy to solve problems faced in their classrooms. We focus next on the supporting conditions that allowed the development and spread of knowledge about strategies such as re-engagement across the region, arguing that good access to mathematical knowledge and the collaborative, practice-based learning structure provided by lesson study were key factors in knowledge development and spread.

What Is Re-engagement and How Did It Spread?

"Re-engagement" presents strategically selected student work in order to "re-engage" students with a previously explored mathematical idea, elicit their thinking, and help them to examine or re-examine their own thinking and that of other students (Fisher and Keyes 2009; Foster and Poppers 2009). As the examples presented below illustrate, presentation and class discussion of student thinking can help students gain mathematical insights at the same time that teachers gain insights about student thinking. Although teachers in the region did not begin to use the term "re-engagement" until 2006, the roots of this strategy go back to at least 2002, when Linda Fisher (SVMI's director of assessment) participated in a lesson study cycle with mathematics coaches from several districts that were involved in SVMI.

Linda Fisher and the coaches served as lesson study team members for a series of lessons on the area of polygons (Mills College Lesson Study Group 2003) planned and taught by Akihiko Takahashi, who teaches at DePaul University and who served as an elementary teacher in Japan for nineteen years, as well as a teacher educator in Japan. Team members supported this work by collecting data on student thinking during the lessons and by participating in the post-lesson discussions, where the next day's lesson plan was reviewed in light of the data on student thinking. Takahashi taught the series of three consecutive research lessons over a three-day period to a fourth-grade class in SMFCSD. Team members were struck by Takahashi's public use of student work to help students consider key mathematical ideas and misunderstandings. Fisher remarked that Takahashi "slowed down" the lesson and probed student thinking, a process the Japanese call "neriage" (Takahashi 2008). For example, at the beginning of the series of lessons, Takahashi asked students to make a 4-inch by 5-inch rectangle on their geoboards, and to copy their solution onto a worksheet. He posted the solutions shown in figure 18.1, and asked students which was correct. As a student came to the board to show the "four spaces" and "five spaces" along the sides of the 4-inch by 5-inch rectangle, Takahashi marked each space with a curved bracket.

Takahashi asked students to explain why the rectangle on the left is not correct, and followed up with another question:

Takahashi: Why [would] somebody make this shape?

Student: Because they counted the number of pegs, instead of the spaces.

Takahashi: So they counted the number of pegs instead of in between the pegs. Let's see. One, two three four. One, two, three, four, five [circles one *"peg" on the drawing with each count, going down the side and then across the bottom of the rectangle*]. OK, so it's very important to count in between to make a shape like this. Do you agree with that?

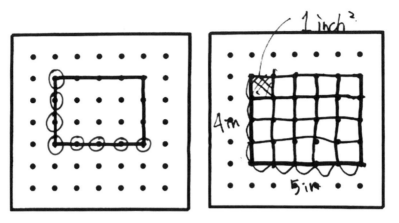

Fig. 18.1. Correct and incorrect examples of 4 × 5 rectangle on board

Likewise, in a discussion designed to connect two different methods for computing the area of a complex figure, Takahashi asked students to compare a strategy that found the area of a 2 × 5 rectangle by adding the area of each row (5 + 5 = 10) to a solution that found the area of the 2 × 5 rectangle using multiplication.

Takahashi: [*To class*] She said this can be five times two. Do you agree with that? Where is five? Where is two? . . .

During the following school year (2003–04) Fisher formed a lesson study team with several network mathematics coaches. The team devoted much of its lesson study time to "pushing on [our own understanding of] the mathematics"—for example, solving proportional reasoning problems, analyzing student work, and reading written resources related to proportional reasoning (including Lo, Watanabe, and Cai [2004]; Singapore Ministry of Education [1994]). Fisher noted that studying the mathematics enabled them to recognize and choose examples of student thinking to discuss with the whole class:

> Playing around and struggling with that mathematics . . . eventually led us to having a really clear idea . . . so when you look at student work you just know . . . You don't have to be doing that thinking on your feet thing that most teachers are doing. It's . . . "Oh yeah, that's an example of this. This is an example of something else." So that as you walked around the room you just see what those examples were that you wanted to pull out . . . But I think it was our own personal struggles that led to that clarity.

In May 2004, three members of the coaches' team co-taught a research lesson centered on the NAEP problem shown in figure 18.2 to a class in Palo Alto Unified School District (PAUSD).

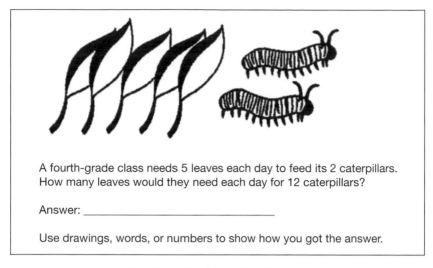

A fourth-grade class needs 5 leaves each day to feed its 2 caterpillars. How many leaves would they need each day for 12 caterpillars?

Answer: _____

Use drawings, words, or numbers to show how you got the answer.

**Fig. 18.2. The Caterpillar Problem
(NAEP 1996, Grade 4, reproduced in Kenney, Lindquist, and Heffernan 2002)**

Figure 18.3 shows some of the student solutions generated by that lesson. The team decided to include these solutions in the revised lesson plan they prepared for a large public research lesson, after the following discussion:

> And remember when Takahashi did it. . . . For each of his problems, he has … that little page about anticipated student responses? I think it would be really [good] for us to put together … one page of … anticipated responses. So that going around the room you could actually tally up how many people are doing each strategy. And then if a, a new one comes up, you only have to find out that one. . . . I think it's possible now that we have some student work to actually do that.

Another team member added, "If there's one that didn't come up that you wanted, you could just … [present it saying] I saw this in another class."

In May 2004, Fisher taught the public research lesson at the Tech Museum of Innovation, attended by about forty math coaches and teachers from the SVMI network. As she collected students' solutions to the caterpillar problem, Fisher told the students, "A really good friend of mine says that the real lesson starts *after* you solve the problem, when you start to think about how other people have solved the problem. I'd … like to start up here with the way Diego solved the problem." Fisher then supported a classwide discussion of five different solution methods used by the students. When students described their thinking to the class in words like "for every two caterpillars they eat five leaves," she encouraged elaboration by asking "Where's the 2 and where's the 5?" and by asking students to point to these spots in their diagram as well as in other diagrams. The five solution methods were thus used to re-engage students in thinking about the mathematics of this proportional reasoning problem from different perspectives (such as unit rate and multiplicative relationships) and to connect these different mathematical ideas.

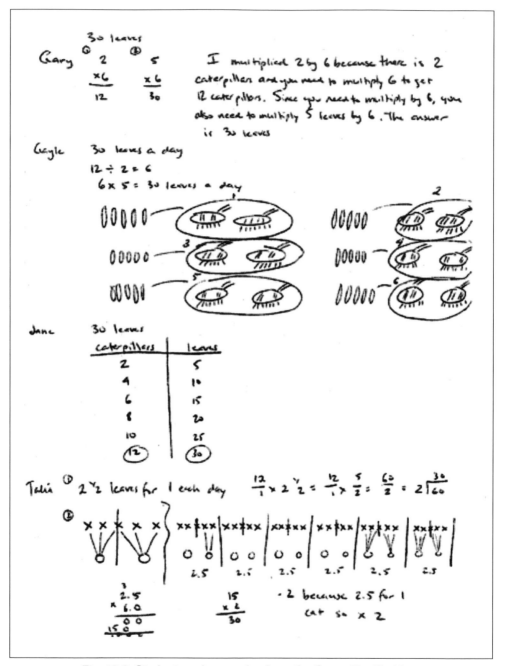

Fig. 18.3. Student work examples from the Caterpillar Problem
included in lesson plan for public research lesson

The mathematics teachers from Willard Middle School (BUSD) viewed Fisher's research lesson and were struck by how well their own students (who had been bussed in for the lesson) explained their ideas during the research lesson. Over the next three years, the Willard group began to add to their own research lessons more opportunities for students to explain their thinking and to make connections between different solution strategies. Figure 18.4 shows the board from a research lesson taught at Willard Middle School in January 2007; its left side records student predictions and explanations about the height a ball would bounce from different dropping points. The planning leading up to the lesson included extensive discussion of the mathematics of proportional reasoning, the solution strategies students might come up with (such as unit rate, scale factor and ratio), the importance of each, and the connections among them. The lesson used re-engagement, introducing examples of student thinking from previous lessons. The lesson also asked students to present their ideas to each other in posters and at the board, and the teacher organized student thinking and data as shown. However, as the post-lesson discussion reveals, teachers saw the need for even more thorough planning of how to present and discuss student thinking:

Teacher 2:   I think [the teacher] did a great job of getting all the student thinking and trying to make a quick gloss ... on that far left-hand side of the board [fig. 18.4], but I think we as a planning group need to think more about ... as a next step, how we can display those strategies in a different way, or more clearly ... either on poster paper or bigger, somehow differently, so ... that kids who didn't use that one ... could benefit from it, and see it, and . . . make connections between these strategies. . . . Because ... there were a lot of strategies there, and I'm not sure how many groups understood or benefited from the strategies that they didn't use in their ... particular group.

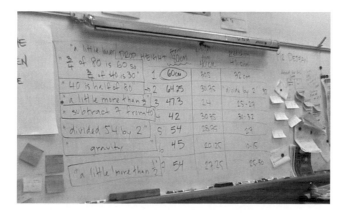

Fig. 18.4. Board from research lesson, Willard School

Meanwhile, Fisher continued to work in lesson study groups and action research groups to examine strategies for choosing, presenting, and discussing student work. These groups tried introducing selected student work (typically from the prior lesson) at the beginning of a lesson so that students could revisit and discuss an important mathematical issue. This strategy, which they referred to as "re-engagement," became a heavy focus of work in the middle school consortium. The Willard lesson study group partnered for an exchange lesson with one of these consortium groups, and they were impressed with the power of re-engagement—for example, the possibility of reintroducing selected student work at the beginning of a lesson so that students could, in the words of Jake Disston:

> Re-engage with these fragile ideas in ways that help them solidify them, and help them weigh the sensibility of them. . . . Instead of just that first pass where they kind of grab something, and it either works or it doesn't work … and then you move on.

In a June 2008 research lesson, Willard School teachers used re-engagement to address a particular mathematical concern of their group. Teachers had noticed that some students who used proportional reasoning when problem quantities were easily multiplied reverted to incorrect additive reasoning when confronted with a non-integer scale factor such as 2.5. Figure 18.5 shows the (recopied) student work used to introduce the lesson. The lesson began by asking the class to describe the thinking behind each piece of work, to explore how each type of thinking would influence the shape of the enlargement, and then to apply each type of thinking to a subsequent problem about lemonade solutions, predicting how each type of thinking would affect the "lemoniness" of the solution. In contrast to the bouncing balls lesson, which included many on-the-fly decisions about how to use student thinking, this research lesson planned in advance how the additive and multiplicative thinking from the poster problem would be used to support discussion of the lemonade problem.

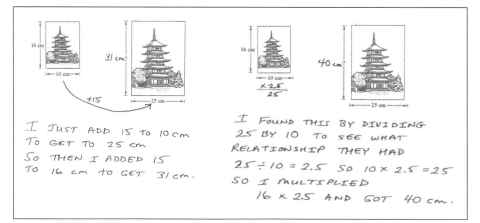

Fig. 18.5. Poster task examples used to introduce Willard research lesson (Task provided by the Mathematics Assessment Collaborative, Noyce Foundation)

Teachers in other districts, as well, were applying re-engagement to various mathematical topics. Many teachers had seen the re-engagement strategy demonstrated in public research lessons taught during annual meetings of the regional lesson study network. For example, SMFCSD teacher Mareva Godfrey (who observed Fisher's group at the January 2008 regional network meeting) tried out re-engagement in her teaching back in her own classroom. She wanted to try it because:

> Typically, our math ... program asks us to invite students to come up and share different algorithms they used, *but for the same answer, the correct one* [emphasis added]. This [using re-engagement] gave me an opportunity to look through the student papers [after the lesson], group answers, whether right or wrong, and look for patterns in misconceptions. Then the students addressed the misconceptions through the discussion. Of course, the correct answer and the different ways of solving the problem were also discussed.

She noted that the re-engagement strategy gave her time between lessons to analyze student work, identify important misunderstandings, and rewrite solutions in her own writing:

> Even though some of the kids recognized their particular answers and algorithms, because the charts up for discussion were in my own writing, [students] seemed emotionally distanced from what I presented and were very engaged in solving the mystery "What was this student thinking?"

Godfrey brought the re-engagement strategy to her lesson study group within SMFCSD, which added it to a subsequent research lesson on bar graphing. Godfrey and teammates also gathered data on the impact of the re-engagement strategy. When students were handed back their own papers, after the re-engagement phase of the lesson, twenty-eight of the thirty students in the class changed their numerical answer and/or added to their explanation, suggesting that they had improved understanding.

The re-engagement strategy also spread to Emery Unified School District (EUSD) through coaches in the SVMI network. Four teachers at Anna Yates School experimented with re-engagement in a wide range of lessons (regular classroom lessons as well as research lessons), using re-engagement to compare strategies to subtract multidigit numbers, to consider what constitutes a good mathematical argument, to revisit responses to difficult open-ended assessment items, and to explore various issues within language arts instruction. Anna Yates teachers observed students during research lessons to see whether and how students used the information from the re-engagement phase of the lesson to revise their own work during the latter part of the lesson, data which convinced them of the useful "dynamic that happens when students look at their own work and that of friends. . . . It's easier for them to get it when other students say it."

In summary, educators in lesson study groups across the San Francisco-Silicon Valley region actively developed, refined, and spread a solution to a persistent problem of practice—how to make student thinking visible and available for discussion. Teachers adapted re-engagement to different topics, different grade levels, and different purposes (such as making links between solution strategies or analyzing misunderstandings). Teachers observed students and studied their work to see whether and how the "re-engagement" phase of the lesson affected students' development of the target mathematical ideas, for example, whether students could explain how additive and multiplicative thinking would play out in a new problem. Practitioners took initiative in using and spreading re-engagement, building in useful features (such as recopying student work in the teacher's handwriting, collecting data on how many students revised their thinking, and pre-planning the choice of student work and flow of ideas around it). Re-engagement enabled teachers to use, within the rapid flow of classroom instruction, knowledge about student thinking and mathematics developed in professional development settings outside the classroom. Re-engagement also spread from mathematics to language arts and from the SVMI network to other lesson study groups.

The Context for Development and Spread of Re-engagement

The development and spread of re-engagement illustrates the potential of a regional professional learning community in which teachers take substantial, ongoing initiative to improve practice through design, teaching, and analysis of research lessons. Foundation personnel, funds, and technical assistance provided an infrastructure for professional learning that included opportunities for lesson study (within and across districts) as well as other opportunities to learn about mathematics and its teaching (such as coding of the MARS assessment). However, this is not a case in which teachers implemented reforms defined by outsiders. Rather, teachers actively worked together and drew on many kinds of knowledge resources (including one another's practice) to develop, refine, and disseminate their collective knowledge about mathematics teaching.

What conditions allowed the development and spread of knowledge about teaching in this region? Four aspects of the case are noteworthy.

1. Teachers had access to high-quality mathematical and pedagogical knowledge through SVMI. Recall, for example, how members of the SVMI's coaches lesson study group worked with Fisher to "push on the mathematics" of proportional reasoning, enabling them to identify the specific student thinking to be noticed and developed during lessons.

2. Lesson study allowed *practice-based* sharing and development of knowledge. Teachers could see what colleagues meant by ideas like "eliciting student thinking," providing a common referent for development of knowledge about teaching. In an early lesson study group with coaches, Fisher was quite surprised to see how her teammates implemented teaching strategies discussed by the team, leading her to remark on "the difference between talking about an idea or strategy and seeing it in action." Lesson study enabled teachers to gain knowledge about their students' mathematical thinking during lessons, and also to bring knowledge from their mathematics professional learning outside the

classroom (such as scoring MARS assessments) into live classroom instruction. (It also allowed SVMI personnel to see how ideas from professional development were translated into classroom instruction, providing valuable formative feedback on SVMI's professional development initiatives.)

3. Lesson study and daily practice were both *ongoing*, so that teachers could progressively refine their ideas through repeated trials in practice. For example, Willard teachers initially found that some students did not take advantage of the ideas summarized on the board (see fig. 18.4), and this prompted teachers, during the next lesson study cycle, to preplan how student ideas would be presented visually and selected for discussion.

4. A strong *collaborative* infrastructure was provided by SVMI and by lesson study. SVMI personnel collaborated on lesson study teams and put their own teaching out for observation and discussion during research lessons, creating an unusually level playing field for learning between the providers and consumers of professional development. Lesson study protocols and norms emphasized collaborative inquiry into student thinking rather than evaluation of one another as teachers (Lewis 2002). Lesson study teams provided opportunities to collaborate with, learn from, and be challenged by colleagues from other schools and districts, thereby allowing teachers to become members of a broader professional community dedicated to instructional improvement. Membership in a valued community provides an important source of motivation and identity (Solomon et al. 2000; Wenger 1999), and participation in lesson study can increase the perceived effectiveness of learning with colleagues (Perry et al. 2009). The close observation of students at the heart of lesson study is also likely to build teachers' motivation. Prior research has shown that students are a powerful motivator of teachers' improvement efforts (McLaughlin and Talbert 1993), and lesson study enables teachers to see instruction through the eyes of students as well as through the diverse lenses of their colleagues.

Teachers' leadership in building knowledge was supported, sustained, and amplified, we posit, through "virtuous cycles" in which teachers experienced greater efficacy in the classroom as they used strategies like re-engagement, stronger identification as members of a professional community devoted to improving practice, and increased motivation to further build and spread knowledge.

Working collaboratively to solve problems of teaching, particularly under conditions of good access to mathematical and pedagogical knowledge, may help teachers strengthen both their collective sense of agency and their participation and identification with a community of practitioners dedicated to improvement of instruction.

......

This material is based upon research supported by the National Science Foundation under grants REC 9814967 and REC 0207259, and on work supported by the Noyce Foundation, the Dirk and Charlene Kabcenell Foundation, Santa Clara Valley Math Project, Silicon

Valley Community Foundation, and contributions from forty public school districts in the greater San Francisco Bay Area. Any opinions, findings, and conclusions or recommendations expressed in this publication are those of the authors and do not necessarily reflect the views of the National Science Foundation or other funders.

REFERENCES

Ball, Deborah L., and Hyman Bass. "Making Believe: The Collective Construction of Public Mathematical Knowledge in the Elementary Classroom." In *Yearbook of the National Society for the Study of Education, Constructivism in Education*, edited by D. C. Phillips. Chicago: University of Chicago Press, 2000.

Boyd, Sally E., Eric R. Banilower, Joan D. Pasley, and Iris R. Weiss. *Progress and Pitfalls: A Cross-Site Look at Local Systemic Change through Teacher Enhancement*. Chapel Hill, N.C.: Horizon Research, Inc., 2003.

Fernandez, Clea, and Makoto Yoshida. *Lesson Study: A Japanese Approach to Improving Instruction through School-Based Teacher Development*. Mahwah, N.J.: Lawrence Erlbaum Associates, 2004.

Fisher, Linda, and Sally Keyes. "We Have What You Want: High-Quality Mathematics Tasks That Improve Learning." Powerpoint presented at Imagine: Mathematics Assessment for Learning (A Convening of Practitioners and Partners) Conference (Session III), Chicago, Bill & Melinda Gates Foundation, 2009.

Fisler, Jennifer L., and William A. Firestone. "Teacher Learning in a School-University Partnership: Exploring the Role of Social Trust and Teaching Efficacy Beliefs." *Teachers College Record* 108, no. 6 (June 2006): 1155–85.

Foster, David, and Audrey Poppers. *Using Formative Assessment to Drive Learning. The Silicon Valley Mathematics Initiative: A Twelve-Year Research and Development Project*. Palo Alto, Calif.: The Noyce Foundation, 2009.

Gorman, Jane, June Mark, and Johannah Nikula. *A Mathematics Leader's Guide to Lesson Study in Practice*. Portsmouth, N.H.: Heinemann, 2010.

Kenney, Patricia A., Mary M. Lindquist, and Cristina L. Heffernan. "Butterflies and Caterpillars: Multiplicative and Proportional Reasoning in the Early Grades." In *Making Sense of Fractions, Ratios, and Proportions*, 2002 NCTM Yearbook, edited by Bonnie Litwiller and George Bright, pp. 87–99. Reston, Va.: NCTM, 2002.

Lesh, Richard A., Mark Hoover, Bonnie Hole, Anthony E. Kelly, and Thomas R. Post. "Principles for Developing Thought-Revealing Activities for Students and Teachers." In *Handbook of Research Design in Mathematics and Science Education*, edited by Anthony E. Kelly and Richard A. Lesh, pp. 591–646. Mahwah, N.J.: Lawrence Erlbaum Associates, 2000.

Lewis, Catherine. *Lesson Study: A Handbook of Teacher-Led Instructional Change*. Philadelphia: Research for Better Schools, 2002.

Lewis, Catherine, Rebecca Perry, and Jackie Hurd. "Improving Mathematics Instruction through Lesson Study: A Theoretical Model and North American Case." *Journal of Mathematics Teacher Education* 12, no. 4 (August 2009): 285–304.

Lewis, Catherine, Rebecca Perry, Jackie Hurd, and Mary Pat O'Connell. "Lesson Study Comes of Age in North America." *Phi Delta Kappan* 88, no. 4 (December 2006): 273–81.

Lewis, Catherine, and Ineko Tsuchida. "Planned Educational Change in Japan: The Case of Elementary Science Instruction." *Journal Of Educational Policy* 12, no. 5 (September 1997): 313–31.

Lo, Jane-Jane, Tad Watanabe, and Jinfa Cai. "Developing Ratio Concepts: An Asian Perspective." *Mathematics Teaching in Middle School* 9, no. 7 (March 2004): 362–67.

McLaughlin, Milbrey W., and Joan E. Talbert. *Contexts That Matter for Teaching and Learning: Strategic Opportunities for Meeting the Nation's Educational Goals.* Stanford University: Center for Research on the Context of Secondary School Teaching. Original edition. Stanford University, 1993.

Mills College Lesson Study Group. *Can You Find the Area? Three Mathematics Research Lessons.* Oakland, Calif. DVD, 2003.

Perry, Rebecca, and Catherine Lewis. "Building Demand for Research through Lesson Study." In *Research and Practice in Education: Building Alliances, Bridging the Divide,* edited by Cynthia E. Coburn and Mary K. Stein. Lanham, Md.: Rowman & Littlefield, 2010.

Perry, Rebecca, Catherine Lewis, Shelley Friedkin, and Elizabeth Baker. "Teachers' Knowledge Development during Lesson Study: Impact of Toolkit-Supported Lesson Study on Teachers' Knowledge of Mathematics for Teaching." Paper presented at the Annual Meeting of the American Educational Research Association, San Diego, March 2009.

Singapore Ministry of Education. *Primary Mathematics 5A, 6A. Teacher's Guide.* Ministry of Education, Curriculum Planning and Development Division. Singapore: Federal Publications, 1994.

Solomon, Daniel, Victor Battistich, Marilyn Watson, Eric Schaps, and Catherine Lewis. "A Six-District Study of Educational Change: Direct and Mediated Effects of the Child Development Project." *Social Psychology of Education* 4, no. 1 (March 2000): 3–51.

Takahashi, Akihiko. "Beyond Show and Tell: Neriage for Teaching through Problem-Solving: Ideas from Japanese Problem-Solving Approaches for Teaching Mathematics." Paper presented at the Annual Conference of the International Congress for the Psychology of Mathematics Education, Mexico City, July 2008.

Wang-Iverson, Patsy, and Makoto Yoshida. *Building Our Understanding of Lesson Study.* Philadelphia: Research for Better Schools, 2005.

Wenger, Etienne. *Communities of Practice: Learning, Meaning, and Identity.* New York: Cambridge University Press, 1999.

CHAPTER 19

Negotiating Cultural Meanings

A Large-Scale Collaboration among Mathematicians, Mathematics Teacher Educators, and Teachers

Gladis Kersaint
Sandra Berger

COLLABORATION, A RECURSIVE PROCESS where people work together to achieve common goals, inherently brings together individuals with different backgrounds and expertise to share knowledge, learn from each other, and build consensus. When designing a model for professional collaboration, it is essential to be aware of the cultures from which the partners will come and, if the collaborative effort is to succeed, to acknowledge and respect the cultural differences of the individuals. *Culture,* as used here, is not defined as a fixed set of norms tied to ethnicity. Instead of this static definition, culture is defined here in the context of cultural funds of knowledge, or "historically accumulated and culturally developed bodies of knowledge and skills essential for household or individual functioning and well-being" (González, Moll, and Amanti 2005, p. 72). This perspective on culture as funds of knowledge takes into consideration an individual's integrated system of experiences (home, school, and work), values, views, and behaviors—all of these influence the meaning that people attribute to their own roles and the lens from which they view and gauge the roles and expectations of others within any collaborative effort. As part of a large-scale statewide effort to provide professional development to enhance teachers' content knowledge, we found that rather than identifying a dominant culture to which others must subscribe, it was important to create experiences that were unique to the partnership. The overall goal is to allow individuals to reveal their own and honor others' cultural perspectives. These experiences provide a foundation upon which to frame continued exchanges and to facilitate the negotiation of meaning.

This chapter discusses the opportunities and challenges that presented themselves when individuals with a variety of cultural funds of knowledge (e.g., disciplinary, institutional, and home) were brought together to design and deliver a professional development program to enhance K–12 teachers' mathematics content knowledge. To provide a context for this discussion, we first describe the project that was the impetus for the collaboration. Then we discuss key areas in which the various cultures influenced the nature of the collaboration. We conclude by reflecting on how the need to negotiate meaning, compromise, reach consensus, and deliver the professional development program provided opportunities for finding common ground and establishing shared norms specific to the project.

Project Overview

In 2007, Florida revised its mathematics standards. The state used the NCTM (2006) *Curriculum Focal Points* as a basis for revising the K–8 mathematics standards, and it restructured the 9–12 mathematics standards into Bodies of Knowledge (e.g., algebra) (Florida Department of Education 2007). These revised standards, particularly at the K–8 level, represented a significant departure from the previous state standards by decreasing the number of benchmarks to be taught at each grade level, in some cases from eighty-seven to seventeen. These changes to the curriculum had implications for teacher knowledge and classroom instruction. For instance, given the narrowed scope of the curriculum, teachers now have time to develop an understanding of the content with greater depth and to use a variety of instructional strategies, including inquiry-based practices. This resulted in a need to provide professional development that would:

- Examine the nature of the curriculum (What is the intent of the revised standards? What are student expectations?)

- Enhance teacher content knowledge (How do teachers support student learning of mathematics in depth if they have weaknesses in their own knowledge? What are ways to address gaps in teachers' content knowledge? How do we help teachers develop conceptual understanding of the subject matter?)

- Enhance pedagogical content knowledge (How do we support teachers so that they teach in ways that build on students' prior understandings, use inquiry-based instructional practices, facilitate student learning, and use the additional time they have been provided appropriately?)

To address this need, Florida awarded a grant through its Mathematics and Science Partnership (MSP) program (United States Department of Education 2010). The awarded project, Florida PROMiSE (*P*artnership to *R*ejuvenate and *O*ptimize *M*athematics and *S*cience *E*ducation), is a statewide project that involves four universities, four large school districts, and thirty-six small, rural high-needs school districts represented by three educational consortia. Key individuals were recruited from partner institutions; these individuals came with different disciplinary backgrounds, teaching experiences (e.g., K–12 and/or university), and perspectives (see table 19.1). Each individual brought very different institutional cultural and subcultural funds of knowledge to the project. By design, there was the K–12 public school culture, the university college of education culture,

Table 19.1

Identifying key collaborators to engage in the content project

Professional Group	Background
Teachers	School districts identified master teachers who could directly support the project efforts and partner with mathematicians in the development and implementation of content-specific professional development. A criterion for selection included prior leadership role in developing and delivering professional development in the local setting.
Mathematicians	A core requirement of the MSP project is the direct involvement of content faculty from the College of Arts and Sciences. Mathematicians were recruited from faculty at partner institutions. Some of these individuals had limited or no previous experiences working with K–12 teachers.
Teacher Educators	Teacher educators were recruited from partner institutions to facilitate and to serve as a bridge between teachers and mathematicians.

and the university college of arts and sciences culture. Institutionally, the unique cultures of three large research universities greatly added to the dynamics of the collaboration. The public school institutional culture encompassed the rural, suburban, and urban differences as well as some corresponding differences in professional development policies, customs, and expectations. Individually, mathematicians brought perspectives from their experiences as mathematics learners shaped by their educational experiences in their country of origin (the majority were not educated in the U.S.) and their experiences as mathematics instructors in the United States. Mathematics teacher education faculty brought with them the pedagogical perspectives of the mathematics education community, perspectives about teacher development and education, and knowledge about how mathematics should be taught and learned. Among other things, K–12 teacher leaders brought classroom insights from their own teaching experiences and from providing professional development in their local contexts. The result was a multilayered collaborative of regional teams within a larger statewide team, representing a myriad of cultural funds of knowledge, both institutional and human.

PROMiSE utilized the expertise of each partner to reach the common goal of improved teacher content knowledge in support of student achievement in mathematics. PROMiSE identified the following principles to guide the nature of its work:

- In order to foster students' conceptual understanding in the ways expected by the revised Florida standards, teachers must have a robust and flexible knowledge of the subject matter. This should be generated through experiences that allow participants to make sense of and reason about mathematics in ways that question current understandings, identify misconceptions, build upon prior knowledge, and make connections among mathematics ideas.

- Teachers should engage in the type of learning that is expected of their students—inquiry-based mathematics problem solving and investigation activities and tasks.

- Professional development, guided by content experts (university content and teacher education partnered with master teachers), should provide teachers with opportunities to learn subject matter through an inquiry process.

Four mathematics institutes were developed and offered to teachers in multiple settings across the state (see table 19.2). A statewide institute design team (e.g., algebra), was made up of multiple regional delivery teams that included a mathematician, two teachers, and, when available, a mathematics teacher educator. The team met to reach consensus on what topics to address in a two-week long institute and how to best address those topics. Once decisions were made about the overall nature of the institute, regional teams, supported by the project leaders, were assigned to further develop components of the common institute to be delivered and to circulate the various drafts for feedback from others on the statewide team. After a final draft that represented the intended design of the institute was accepted by the statewide design team, regional teams planned for local implementation.

Table 19.2

Topic, audience, and number of the summer institutes provided

Topic	Audiences	Number of Sessions
Number and Operations	Teachers of grades K–5	2
Rational Number and Proportional Reasoning	Teachers of grades 3–8	4
Geometry	Teachers of grades 3–8	2
Algebra	Teachers of algebra in grades 6–12	5

Bridging Cultural Boundaries: The Need for Shared Understandings

Although mathematicians, mathematics teacher educators, and teachers represent expertise about mathematics education in their various capacities, their cultural funds of knowledge within particular educational contexts frame their understanding of issues that are specific to mathematics and its teaching and learning. Because of these varied perspectives, collaborators must develop a shared understanding of the intended goals and agree on what is needed to attain them. With PROMiSE, we found several concepts that needed to be negotiated because they had different meaning depending on an individual's prior experiences. Below we highlight these key elements. In the sections below, we use the term "educator" to represent a group that includes both teachers and mathematics teacher educators.

What Does It Mean to Be a Team Member?

Individuals with different backgrounds and expertise were brought together because it was assumed they would bring mathematics education experiences upon which the partnership could build and learn. However, it soon became evident that there was not even a shared understanding of what it means to "collaborate" on a common endeavor. There appeared to be a need to establish who was to be considered "the authority" for decision-making purposes. Individuals had been identified to participate because they were leaders of their respective groups and, therefore, all assumed they were *the leader* of the team of which they were part. Some mathematicians thought that, as content experts, they should lead, while some teachers thought that their knowledge of local classrooms and teachers should give them the final say. Mathematics teacher educators assumed that they were uniquely positioned to lead because they could cross both boundaries, but others questioned whether they were sufficiently knowledgeable about either context.

Although individuals from each group were willing to work with each other, an inherent distrust existed as to how each could contribute to achieve the intended goals, considering their different backgrounds. Although it should not have been surprising, some of the statewide design team members were frustrated that their professional peers (i.e., other mathematicians or other educators) did not share common views. Mathematicians from different universities, where they were the recognized leader in their field, tended to discount the contributions of mathematicians from other universities, and teachers from large districts discounted the contributions of teachers from rural districts and vice versa. Much of the frustration was revealed when individuals initially considered being asked to explain or justify a suggested approach as a personal affront and a challenge to their expertise.

The challenge was to get individuals to drop their role of teacher and expert and instead to take on a new role as collaborator and learner. To mediate statewide and regional team discussions, the project leaders had to serve as "cultural brokers." The leaders constantly reiterated that expertise is highly contextualized and that the strength of the design teams was in the different perspectives each member provided. Members were often reminded that they would serve as both a leader and a follower and that through discussions and deliberations each would learn, contribute to the learning of others, and, ultimately, develop a final product that would be enhanced by the contributions of all. In addition, it was important to make explicit that an individual should not assume that his or her knowledge was commonly shared by all, or even by those with the same job title.

As team members became familiar with each other, the nature of the conversations advanced, and team members began to appreciate the knowledge and perspective that others provided. This became particularly evident after the implementation of the first summer institute, when members of each team had an opportunity to witness the impact of tasks and instructional practices suggested by themselves and others.

What Does It Mean to Know Mathematics?

Most people would agree that an understanding of content is critical for teaching, but there is little agreement on what constitutes understanding of the content needed by teachers (see, for example, Ball, Thames, and Phelps [2008]). Institute design teams, therefore, had to make sense of what it means to know mathematics before they could make sense of how teachers need to understand a mathematical concept to support instructional practices. Although one might assume that there would be a predictable difference across the institutional cultures in determining concept meaning (i.e., teachers/mathematicians), there was also a need for negotiation within groups (i.e., educator/educator, mathematician/mathematician). To support their deliberations, institute design teams were provided with resources (e.g., existing professional development programs, curricular materials, and research reports about student and teacher knowledge) that they could use or adapt as they developed a particular institute. As can be expected, individuals gravitated toward resources that supported their particular perspective. In the end none of the teams chose to use or adapt an existing program.

Below we highlight some areas of the extensive discussions that were held across the various institute teams as they designed professional development to enhance teacher content knowledge. Comments should not be construed as representative of the culture of a particular profession (e.g., educators or mathematicians), as there were great variations within each group and design team. However, when appropriate, we highlight comments that represent statements made most often by members of a particular group (e.g., teachers). Also, representative comments are shared as a means to highlight major points of divergence not only within particular institute design teams, but across the various teams.

Inquiry Learning

A tenet of the professional development model we used was that teachers should experience learning mathematics from an *inquiry perspective*. That is, participants would be given opportunities to question their understandings through exploration and investigation, to solve problems in different ways, and to present arguments and question those presented by others, as suggested by the NCTM Process Standards (NCTM 2000). Although it was assumed that inquiry would be more difficult for the mathematicians, who might be more accustomed to delivering well-crafted lectures, notions of inquiry were difficult for many of the educators as well. Throughout the collaborative effort, institute design teams grappled with what is meant by inquiry-based approaches. For example, a mathematician asserted that to ask him to teach in this way was to deny how he learned mathematics, and that it suggested there is no value in developing clear lectures to help people learn. Other participants, including other mathematicians, saw value in approaching mathematics from different perspectives as a means to provide broader access to it. Because of their culturally developed notions of teaching, some collaborators found it difficult to envisage their roles as part of an inquiry-based classroom. For example, those who were most comfortable giving a lecture did not understand how they were expected to facilitate learning. To honor individuals' prior experiences, the project leadership continually reinforced that there is merit in any method that facilitates learning, but that this project is guided by the instructional expectations of Florida's revised mathematics standards.

Depth of Understanding

Institute design team members were asked to design professional development to enhance teachers' depth of content knowledge. The institute topics were not to be limited to those that teachers specifically teach, but should include ones that teachers would need to know and understand in order to facilitate students' depth of understanding. Again, historical cultural funds of knowledge made this a difficult task for design team members who needed time to grapple with their own perception of what is meant by "depth of knowledge" and, subsequently, how that depth of knowledge is developed. For many, the term meant more advanced levels of mathematics rather than a "deeper understanding" of the underpinning of particular concepts. Some design team members believed that in order to develop depth of knowledge, participants would need ample opportunities to practice learned procedures. Others believed that it is important to think about the same topic from multiple perspectives; that is, finding the solution for many systems of equations was more important than modeling situations with systems of equations and being able to conceptualize and interpret the solution set of the system of equations. Even team members who believed in the need to develop conceptual understanding of mathematics could not articulate what was needed to develop conceptual understanding, and in some cases equated the use of concrete materials with conceptual understanding. These views represented the cultural perspectives that had been developed in an individual's own educational experience as either a learner or a teacher. Sometimes an individual had been introduced to the terminology and an instructional strategy but had not had an opportunity to fully find meaning for the words in practice.

Mathematical Focus of the Institute

Some of the collaborators were surprised by the lack of agreement regarding which topics to include in a particular institute, and on what is important to know and understand about that mathematics topic (e.g., algebra). Some mathematicians thought it would be valuable to expose teachers to significant historical developments in mathematics (e.g., estimation of square roots with continued fractions for K–5 teachers); some teachers, on the other hand, believed that these developments were too far removed from their work as teachers to be relevant. Rather than examining how the content would build upon and enhance teachers' prior understanding, initial conversations centered on including an individual's favorite topic or activity without necessarily focusing on the overall goals for their particular institutes. In the algebra institute, for example, a mathematician insisted that no algebra course was complete without a discussion of Bezout's theorem but did not explain why, or identify where it might fit. Likewise, a teacher wanted to emphasize the development of the quadratic formula in a variety of ways but did not articulate its importance or placement within other topics to be addressed in the institute. Others argued that teachers needed to have experience with particular manipulatives (e.g., algebra tiles). These ideas have value, but there was an initial assumption of importance that was based on an individual's cultural funds of knowledge, without recognizing the need to provide rationales to help those with different experiences understand how these topics or approaches would enhance teachers' knowledge or build connections among the other suggested ideas. Overall, identifying topics important to particular institutes was a long reciprocal process that required negotiation and compromise.

Nature of Mathematics Activities

To support notions of inquiry learning, team members were asked to use worthwhile mathematics tasks or investigations to explore mathematical concepts (NCTM 2007). A very fundamental belief difference surfaced during discussions about how mathematics is or ought to be learned. Two major points of contention arose around which should come first—abstract mathematical concepts or experiences with concrete objects, and taught procedures or open-ended problems. Group discussions about representing mathematics included the view that mathematics is inherently abstract and the view that concrete representations are necessarily conceptual. Educators typically asserted that it was important for the participants to experience a concept through exploration (in some cases with concrete tools) prior to being exposed to it formally (i.e., in precise definitions). In contrast, many of the mathematicians asserted that one must understand the abstract in order to apply the mathematics as a concrete concept. In fact, when asked to present a topic in a less abstract way, the response from one mathematician was that being less abstract would not be staying "honest" to the discipline. For example, in developing the concept of transformation, a mathematician argued that teachers needed experiences with the symbolic representation of a rotation (i.e., $R_{0,180}(X) = X$) before they could apply it (e.g., use of patty paper to explore what happens). To insist that it be done any other way contradicted what for him had been an accumulated skill essential to his individual functioning. There was also disagreement on whether to provide clear and precise definitions before exploring mathematics or to explore mathematics ideas first and then formalize and clarify conjectures and definitions. Again, few team members had ever experienced inquiry strategies as a student of the discipline, so there was no historically accumulated body of knowledge from which to draw.

The tasks initially developed by some collaborators, both mathematicians and mathematics teacher educators, indicated their misconception that conceptual understanding develops automatically when one engages in investigations or uses concrete materials. In these cases, the designed task/activity did not necessarily clarify or develop the underlying mathematics concept. Tasks were proposed as either "fun" ways to engage participants in practicing a mathematical concept that had been previously explained, or as a "fun" activity in which a mathematician could "find some mathematics." Some activities promoted social interactions but did not support mathematics learning. Some of the collaborators suggested that inquiry approaches were being used if participants were asked to talk or contribute to a discussion as part of an interactive lecture, usually with no consideration of the type of questions asked. For example, one "activity" was for the group to "find the next perfect number after 6." Some educators suggested that a hallmark of inquiry-based approaches is the use of *unguided* explorations, the assumption being that if participants are provided with the needed tools they can make sense of the mathematics. It was assumed, for example, that participants could model operations with base-5 blocks if they were simply provided the blocks without discussion. Still others, both mathematics teacher educators and mathematicians, saw the use of technology as a crutch that allowed the "machine" to generate results instead of students learning *real* mathematics. Typically, individuals with strong opinions about the inappropriate use of manipulatives and technology had limited experiences with them.

Overall, early design team discussions often focused on what activity to do rather than on which ones would develop mathematical meaning. As individuals shared ideas, and defended or refuted the ideas of others, the cultural basis for their assumptions became apparent. Many of the shared ideas were based on an individual's own experiences as a student, a teacher, or a professional development provider and, as a result, the potential benefits of alternatives had not been considered. In other cases, individuals have taken ownership of ideas suggested by the broader education community (e.g., the use of manipulatives) but had limited conceptions about how they should be best used to support mathematical learning.

Cultural Uses of Language

The different vocabulary used by design team members was an expected challenge, considering the various funds of knowledge represented. Words used in one educational setting were often unfamiliar or used differently in the other. Though we expected language to be a challenge, the actual language that was problematic took some surprising turns. For example, a research mathematician reacted indignantly to the suggestion that teachers can be led to "discover" a particular mathematics concept, and stated, "If it were so easy to discover mathematics, it wouldn't have taken the Greeks two thousand years." We also encountered a surprisingly prevalent assumption that teachers and mathematicians spoke about mathematics in different ways. A mathematical error made by a teacher (multiplying by 10 instead of multiplying by 10/10) was defended by a mathematician as "That's how teachers talk"—as opposed to mathematicians, who would use the language of mathematics in a more precise manner. In other cases, phrases were used that were unfamiliar (e.g., "Why do you all call that scientific notation? Isn't it just a notation?" said a mathematician). There was a continual need for design team members to clarify the intended meaning so that others could understand their point and, if needed, share possible alternative meanings.

Negotiating Meaning: Finding Common Ground

The scale and multilayered nature of the partnership produced the need for many compromises to enable the work of the project to progress. Given the various culturally laden perspectives of team members, it was important to ensure that compromises supported the intended goals of PROMiSE rather than simply eliminate the need for difficult discussions. Here we share aspects of the work that helped teams find common ground.

1. **Openness to other viewpoints.** In many cases, the challenges were resolved as team members got to know each other and became more willing to accept that there might be alternative approaches that were not part of their accumulated bodies of knowledge or experiences. Ultimately, the shared norms were the result of openness to varying viewpoints in an effort to support the project goals. This required everyone, including the project leaders, to recognize that contributions by team members provided opportunities to learn from each other and, through the collaborative efforts, to develop a better understanding of individuals from various disciplinary, institutional, and home cultures who were all interested in a similar goal: enhanced teacher knowledge and student learning.

2. **Time to build relationships.** The challenges inherent in bringing together such a wide array of cultural funds of knowledge required a variety of negotiation approaches. Many of the challenges that were identified early were worked out with continued communication and familiarity. As team members got more comfortable with each other, they became more comfortable asking for and giving clarification. Other challenges were overcome by relying on the tenets of the project to guide the members' work. When disagreements occurred, team members were asked to determine the extent to which the suggested idea supported the tenets of the project. Still other challenges were dealt with by design team members choosing to go beyond their comfort zone by utilizing strategies, tasks, and approaches suggested by others—mathematicians leading activities and teachers leading advanced mathematics discussions.

3. **Blending perspectives.** Given the amount of time provided to engage in this work, we did not attempt to alter the various perspectives and views held by team members. Instead, we honored and blended the suggestions of the differing design team members. For example, the order of instruction (i.e., abstract vs. concrete) was handled in different ways in different situations. As one result, many of the mathematicians reported seeing the potential for allowing participants to engage with concepts prior to providing formal definitions or symbolism. Favorite topics and tasks were incorporated with the caveat that there be a strong rationale and connection to the institute topics, and any "favorite" that didn't fit was either dropped or handled as an "optional" topic. The issue of what content teachers could handle was settled with a blending of the advanced mathematics with the topics that teachers are responsible for teaching; many

educators were surprised at teachers' level of engagement with challenging and advanced mathematical tasks. In fact, many teachers reported that they wanted more challenging tasks to complete. They enjoyed the opportunity to simply talk mathematics. In blending inquiry with "interactive lecture," a term used by mathematicians to indicate a dialog with students during a lecture, team members focused on using higher-order questioning techniques and allowing teachers to explore what they would do with the mathematics before "showing" them *the* way to do it.

4. **Continued focus on important issues.** The nature of the challenges evolved over the three-year project. While some mathematicians and educators broadened their perspective of what it means to know and do mathematics, the need for dialog continues. During the project, discussions about the development of depth of understanding moved away from simply providing more advanced mathematics to a focus on deepening understandings. For many, memorization of prescribed algorithms is no longer seen as more efficient than some invented algorithms or processes. Also, there is a growing awareness of the role of investigations and meaning-making in developing mathematical concepts. And most important, our basic tenet of inquiry learning began to coalesce as a formal concept among team members.

5. **Reflecting on implementation.** Because common institutes were implemented by various teams in anywhere from two to five different locations, every team had to deliver aspects of the institutes that they themselves did not design; this required them to make sense of the task along with the recommended pedagogical approaches for using the task. After the first round of implementation, teams who taught a common institute came together to share successes and challenges. Teams learned that what worked well for one team might not have worked as well for another, providing an opportunity to examine the underlying causes for the differences in implementation. In many cases, these differences were linked to pedagogical approaches. In other cases, individuals adapted the mathematics activities developed by others to make them more suitable to their preferred instructional approaches, either enhancing or losing the intent of the original designer.

Through implementation, some collaborators were able to see for the first time the intent or "genius" of the other collaborators, and others readily admitted that they had underestimated the possibilities. A major success for all of the design teams is that the project's external evaluators reported statistically significant gains in participant content knowledge in all but one of the institutes during the summer 2009 implementation (Lauman et al. 2009). During the next academic year, all collaborators continued with the project. They collaborated on the revision and enhancement of the institute based on their common experiences and new insights. The revised institutes were delivered in summer 2010. A measure of the project's success is that statistically significant gains in teacher content knowledge were obtained in all of the institutes delivered in summer 2010 (Slaughter 2010).

Summary

The design and implementation process of this project enabled collaborators to explore approaches learned from others, do and see things differently, and appreciate what they might not have fully understood during the design phase. Engaging in and discussing common experiences provided collaborators with opportunities to construct meaning and develop common understandings. The design process was messy and uncomfortable for many, as could be expected with their different experientially accumulated cultural bodies of knowledge. This type of collaboration was new to most of the team members, and many of their initial reactions were based on their prior culture as defined earlier. However, after the implementation of the summer 2009 institutes, they had a common set of experiences upon which to reflect and use as a basis of continued conversation, allowing team members to appreciate each others' contributions and to interact with greater respect for the expertise of others. This set of common experiences, both the discussion and the implementation, allowed the development of shared norms that added to the participants' cultural funds of knowledge, something that will continue to facilitate future exchanges.

REFERENCES

Ball, Deborah L., Mark H. Thames, and Geoffrey Phelps. "Content Knowledge for Teaching: What Makes It Special?" *Journal of Teacher Education* 59 (November/December 2008): 389–407.

González, Norma, Luis C. Moll, and Cathy Amanti. *Funds of Knowledge: Theorizing Practices in Households, Communities, and Classrooms.* Mahwah, N.J.: Lawrence Erlbaum Associates, 2005.

Lauman, Beatrice, Priscilla Carver, Michael Howard, and Joy Frechtling. *Analysis of the Impact of Florida PROMiSE Summer Institute on Teacher Content Knowledge.* Rockland, Md., September 2009. http://www.flpromise.org.

National Council of Teachers of Mathematics (NCTM). *Mathematics Teaching Today: Improving Practice, Improving Student Learning,* edited by Tami S. Martin. Reston, Va.: NCTM, 2007.

Florida Department of Education. *Next Generation Sunshine State Standards.* http://www.floridastandards.org.

Slaughter, Andrew, and Joy Frechtling. *Analysis of the Impact of Florida PROMiSE Summer Institute on Teacher Content Knowledge.* Rockland, Md., December 2010. http://www.flpromise.org.

United States Department of Education. Mathematics and Science Partnership. http://www2.ed.gov/programs/mathsci/index.html.

CHAPTER 20

Powerful Pedagogical Practices Program:
Creating a Catalyst for Moving to a Problem-Based Curriculum

Jamila Q. Riser

> *Why would I waste class time having students offer explanations for how they solve problems? My explanations are after all [always] the best ones. Ultimately, I end up having to go back and clarify or re-explain the solutions to the class anyway. I don't have that kind of time to waste.*

THIS ATTITUDE, AS EXPRESSED BY ONE LOCAL HIGH SCHOOL TEACHER, may all too often be characteristic of the approach to teaching that secondary math teachers exhibit in their classrooms (Boaler 2008). The sentiment suggests that teachers believe that their skill in demonstrating mathematics will be sufficient to engage and enlighten their students. These experiences and all of their professional preparation have led them to believe that successful teachers are those who provide students with well-articulated explanations about *how to do* math.

In the program that will be discussed below, we aimed to inspire a "new normal" for high school mathematics instruction, one rooted in a culture of engagement in problem-based learning. "Problem-Based Learning (PBL) describes a learning environment where problems drive the learning. That is, learning begins with a problem to be solved..." (Roh 2003, p. 1). In a problem-based mathematics classroom, students are asked to engage in a task for which the solution is not known in advance. Ideally, the teacher poses an intriguing and challenging problem, students spend a period of time independently engaging in the task, and then students work collaboratively with other members of their group in order to solve the problem (Boaler 2008). This problem-centered approach to teaching presumes that "solving problems is not only a goal of learning mathematics but also a major means of doing so" (NCTM 2009).

One important classroom norm that permeates the culture of a problem-based learning environment is that students' ideas and conceptions of mathematics are central to the development and progression of mathematical understanding (Hiebert 1997). This view presumes that teachers recognize and demonstrate a belief that students learn best through listening to other students, communicating about their learning by verbalizing and writing about their own thinking, building on each others' ideas, and evaluating the integrity of one another's mathematical arguments.

This perspective of how one learns and applies mathematical knowledge requires shifting from teacher-centric classroom cultures to a classroom culture in which students' ideas and conceptions of the mathematics are inherently valued. For this to happen, teachers must willingly change their beliefs and practices in fundamental ways. According to Jean Piaget, "The essential functions of intelligence consist in understanding and in inventing," and "by telling a student something that could have been discovered by themselves, you rob them of the chance to truly understand it" (Piaget 1970, p. 715).

Powerful Pedagogical Practices First

This chapter details the major characteristics of the Powerful Pedagogical Practices (P^3) program, a statewide strategic effort aimed at transforming the pedagogical practices of Delaware secondary mathematics teachers. The project, funded through a state-administered Mathematics and Science Partnership grant, was a direct and principled response to demands from members of the Delaware Mathematics Coalition (DMC). The revitalized coalition, which represented an alliance between curriculum directors from each of the public and charter school districts in the state, the department of education, and partners from the higher education and business community, met monthly in order to discuss and develop strategies for making significant improvements in the quality of mathematics teaching and learning.

Both state and national assessment measures and anecdotal information collected from district leaders indicated that the state had made significant progress toward realizing its vision for improved mathematics instruction at the elementary and early middle school levels. Sadly, this was in sharp contrast to what was happening in many of the state's high school mathematics classrooms. Coalition members were prepared to "do something dramatic" to improve high school mathematics achievement, which had been stagnant for the past five years, with two out of every five tenth graders consistently failing to meet the state's minimum standards.

In 2007, the DMC developed a five-year strategic plan, including professional development priorities based on statewide achievement data and on the insights gained from the interviews of district mathematics specialists, teachers, and curriculum supervisors conducted by the coalition's director. The most urgent need related to pedagogical support for high school mathematics teachers. In particular, secondary mathematics teachers complained that they felt ill equipped to address the diverse needs of the learners entering their classroom doors. They also expressed concerns about the lack of opportunities to work with their middle school math counterparts. Curriculum directors and mathematics specialists lamented the fact that high school teachers of mathematics too often seemed more focused on the content they could "deliver" and less aware of the pedagogical alternatives for enabling all of their students to be successful in learning that content. "Focus on the pedagogy first" was the overwhelming consensus. "We'll confront issues of content and curriculum later."

This process led to the conception of the Powerful Pedagogical Practices program. During the following three years, the project was one of the coalition's highest professional development priorities. Unique to the P^3 project were—

- the selection and long-term (3+ years) retention of teams of teachers (grades 8–10) from nearly all the Delaware public school districts;

- the important and strategic choice of addressing pedagogy before curriculum; and

- the ongoing engagement of key stakeholders, including high school principals.

Curriculum directors insisted that for significant and sustained changes in high school mathematics classrooms to take place, high school principals must do more than simply "support" their teachers in promoting change—they must become active advocates of change. As full-time participants in the learning community, P^3 leadership would provide opportunities for administrators to work side-by-side with their teachers in order to develop a common vision for highly effective problem-based instruction, deepen their content and pedagogical knowledge, and develop a common language for talking about the teachers' professional practice.

Launching P^3

In April of 2007, the Delaware Mathematics Coalition formally launched the Powerful Pedagogical Practices program with a statewide meeting. Leaders and designers of the P^3 professional development program included the director of the DMC; the director of the University of Delaware's Mathematics & Science Education Resource Center (MSERC); mathematicians and mathematics educators from the University of Delaware, the Coalition's higher education partner; and master teachers on loan to the university. District and school leaders selected their P^3 teachers based on several criteria, including their impression of the teachers' willingness to explore new approaches to teaching and learning, their commitment to improving the quality of their craft, and their ability to energize and influence their peers. Team principals were selected on the basis of their responsibility for supervising teachers within the mathematics department. Several months later, the first cohort of P^3 secondary mathematics teachers and high school administrators chosen from nearly every district in the state gathered for a multiday summer institute.

District teams, which included grades 8–10 mathematics teachers, a high school administrator, and the district mathematics specialist, continued to meet each summer as well as on a monthly basis throughout the school year for the next three years. Three-quarters of the P^3 participants consistently attended the professional development for all three years of the program. The professional development was purposely designed by project leaders to provide ongoing opportunities for the P^3 community to come together to deepen their content knowledge, study and discuss research related to the pedagogical behaviors that permeate a problem-based learning environment, and reflect on their practice. These pedagogical practices had implications for the behaviors of both teachers and their students. Examples of these behaviors included increasing opportunities for students to share strategies with their classmates, promoting the use of multiple strategies, increasing the percentage of time listening and probing student thinking, and increasing the use of descriptive rather than evaluative feedback in order to promote student understanding (Black et. al. 2004).

Realizing Change in High School Mathematics Classrooms

Much of the research on teacher change suggests that for teachers' beliefs and practices to shift, they must be a part of an ongoing community that collaborates to develop and implement common learning goals. Within the high school setting, however, many teachers report that they most often work in isolation. They do not have access to experienced colleagues who can help them with day-to-day problems and share successes and insights (Shulman 2002; Wilson and Berne 1999). Most teachers, even those who work in close proximity, have minimal opportunities to share their ideas about classroom practice with one another (Grossman 1995). Conversely, when teachers and administrators have ongoing opportunities to work together and develop as a community, this promotes conditions for positive change. "By working together in groups to improve instruction, teachers are able to develop a shared language for describing and analyzing classroom teaching, and to teach each other about teaching" (Stigler and Hiebert 1999, p. 123).

As project leaders, we also assert that the most effective way to promote successful transformations in high school teachers' instructional practices is for teachers and their administrators to have sustained opportunities to be immersed in a problem-based learning environment that models the classroom and student interactions they hope to actualize. According to the National Research Council (1989), for significant changes in *how* mathematics is taught to occur, teachers themselves must experience learning mathematics in ways that promote making and testing conjectures as well as engaging in mathematical argumentation.

The Conception of a Shared Vision of a Problem-Based Mathematics Classroom

NCTM's *Principles and Standards for School Mathematics* paints a compelling and vivid vision of a problem-based classroom, one in which teachers are thoughtful problem posers, eager to challenge and engage students in solving complex and intriguing mathematics tasks. In these problem-based classrooms, teachers' actions demonstrate that they value their students' unique ways of thinking and actively promote a collaborative culture that centers on students defending and making sense of one another's ideas (NCTM 2000).

We believe that in order for long-term and sustainable changes in high school mathematics teachers' classrooms to be realized, teachers must come to value this way of learning and see it as a way to increase access to deep and enduring mathematical understandings for all students. Over time, they must develop a clearer vision for what problem-based mathematics classrooms look like, feel like, and sound like. They must be prepared to honestly assess their current practices, have time to articulate their challenges, willingly set goals, develop action plans, and reflect about their progress toward a "new normal" in high school instruction. In the sections that follow, we share three major foci of our work in P^3.

Doing Mathematics Together

Monthly professional development sessions provided P^3 project leaders with opportunities to immerse teachers and administrators in a problem-based learning environment. During the launch portion of the problem-based lesson, P^3 facilitators posed a rich and challenging problem for cohort members to consider, prompted conjectures as appropriate, and then provided a few moments of individual "think time" before participants began working together to compare strategies and build on each other's ways of thinking. According to the *Principles and Standards*, worthwhile tasks are described as "intriguing, with a level of challenge that invites speculation and hard work; such tasks can be approached in more than one way, allowing access to students with varied experiences and prior knowledge" (NCTM 2000, p. 19).

As participants worked on the given task (the investigative phase of the lesson), P^3 professional development leaders circulated around the room listening to conversations, asking open-ended questions (e.g. "How did you arrive at that conclusion? Do you think that *always* works? Do all of you agree with that?") and gathering data for the share out or summarizing activity that would follow. Our experiences with the P^3 cohort members provided us with additional insights about the delicate nature of the timing or "chunking" of the instructional phases of a problem-based lesson. We theorized that during the investigation, when the "buzz" of the conversations reached a peak and then began to subside, was the optimum time to bring the class together for a share out or consolidation of the learning.

The need for a focus on the summary phase of the lesson became increasingly important, as P^3 teachers' journal reflections and classroom video data suggested that teachers struggled with how to effectively orchestrate a mathematical discussion during this portion of the lesson. All too often, teachers rather than the students were doing most of the talking during the summary. Even in cases where students were playing a role in the share out, teachers would find themselves jumping in to clarify the students' explanations or restating the ideas in their own words. Some teachers also shared their frustrations with not being able to get more students involved in questioning the student presenter.

The summarizing portion of the lesson should ideally provide an opportunity for students to build on and gain new knowledge. A compelling example of this can be found in the introduction to Jo Boaler's book *What's Math Got to Do With It?* In it she describes a scene from a high school teacher's classroom: The students are solving a challenging problem about the time it would take a skateboarder to crash into a wall after holding on to a merry-go-around that is spinning, and then letting go at the "two o'clock" position. She stops the class short of anyone solving the problem and asks the students to offer their ideas and let the remainder of the class know where they are stuck. During the share out, students come up to the board, presenting their ideas to their peers, acknowledging their struggles, and building on one another's ideas. After ten minutes, the class had worked together to solve the problem (Boaler 2008). During a visit to Delaware, Professor Boaler shared the video segment of this classroom scene with our P^3 community. Watching the video of Emily Moskam's classroom inspired our vision for a "new normal" in high school instruction in summarizing mathematical concepts.

Through the use of a set of carefully chosen problem-based tasks, professional development facilitators modeled and explicitly addressed the pedagogical challenges of teaching in this unfamiliar environment. Often these problem-based tasks were launched with a "big" problem to be solved and then engaged learners in a series of smaller problems as a means to create access and promote the conceptual basis for understanding the "larger" problem. One such problem, the "birthday" problem, is presented in the *Interactive Mathematics Program* (IMP) series. The critical question used to motivate student interest and intrigue is "What is the probability that two people in this room share the same birthday?" After posing the initial big problem, students are then asked to investigate a series of questions related to the number of people necessary to have a match of the day of the week or month of birth. Using complementary and independent events, students calculate the theoretical probability of each event. While many secondary mathematics teachers have had experiences studying probability in school, teachers in our project willingly voiced concerns that often their knowledge was fragile, and grounded primarily in mathematical experiences that focused on rules and procedures rather than on deeper conceptual understandings.

When teachers engaged in challenging tasks like these, tasks which *required* collaboration to solve them, and then later used them in their classrooms and witnessed how students enthusiastically and successfully demonstrated an understanding of these problems, they not only developed higher expectations for what they believed their students could achieve, but they also began to develop an appetite for using these tasks with their students on a more frequent basis. At an end-of-year district meeting, one P^3 teacher reflected out loud about the "birthday" video that was shared by one of her P^3 colleagues during the year. She said, "I was sitting with John on the day we did that problem at P^3. I witnessed him *struggling* with this problem. When we watched the video of him doing the problem with kids in his classroom . . . what he was able to accomplish with them . . . *that was golden*! That was the real power of P^3!"

End-of-project interview data collected suggested that the experience of *doing math together* "helped them [the participants] get back in touch with what attracted them to math in the first place. Many spoke as if they were renewing a relationship with math by again being learners. Using mathematics to solve different kinds of problems in different ways than many of them had tried before positioned teachers and administrators as students, a perspective from which they could consider what their students would make of similar experiences" (Fifield 2010, p. 5).

Students as Mathematical Thinkers

By developing an increased interest in their students as mathematical thinkers, project leaders believed that, over time, P^3 participants would increase their capacity to facilitate a classroom culture that capitalized on students' ideas and conceptions about the mathematics. One way project leaders fostered this interest was through the use of videos of interviews of students in P^3 teachers' classes. During these interviews, project leaders (serving as the interviewer) would generally begin with an open-ended question such as "Tell me about how you approached this problem." As the student shared his or her ideas with the interviewer, the interviewer would attempt to probe the child's thinking further, asking clarifying questions as necessary. During the professional development, project leaders would pause the video at certain points to engage participants in conversations about student thinking; these might include anticipating what the student will say next, sharing ideas about a follow-up question they would ask the student, diagnosing a particular misconception or error that surfaced, making sense of the student's strategy, or critiquing the nature of the interviewer's questions.

As participants returned to their classrooms and implemented the kind of problem-based instruction they had experienced during the P^3 sessions, they were encouraged to listen more intently to what their students were saying both in small groups and in large group settings, and to make their student thinking public. Journal reflections from project teachers indicated that they were much more conscious of whether they were really listening to their students and asking questions based on their students' ways of thinking, and not inadvertently leading students to *their* preferred strategies. The teachers considered ways to ask questions to "open up" classroom discourse rather than constraining it. For example, participants began to talk about their attempts to use focusing versus leading or funneling questions (Herbel-Eisenmann and Breyfogle 2005). Funneling questions are characteristically based on the teacher's rather than the student's conceptions of a solution path. They may focus too heavily on a set of procedures for arriving at the desired solution, and they often leave the learner without the tools to make important cognitive connections. The use of focusing questions, on the other hand, requires that the teacher listen to students' responses and guide them based on their way of thinking, rather than the way the teacher might approach the task. This awareness about the nature of their questions and whether they were truly capitalizing on their students' unique ways of thinking seemed to grow over time, especially with greater use of teacher classroom video data.

Action Research

While project teachers were engaged in and committed to enhancing their practice, we wanted compelling data related to the impact of P^3 teachers' changes. This data came in the form of increased use of classroom video data and the selection of a specific action research goal. At the end of the first year, teachers were asked to collect a single classroom video episode that would serve as a "baseline" of their practice at the time. The broad goal of "making student thinking" public remained a constant theme throughout the life of the project, and, during the final two years, participants were provided with a select menu of pedagogical behaviors to choose from for their action research. These goals were based on the principles that were both modeled during the professional development and supported by research related to how students learn. The list provided options for teachers at various stages of their pedagogical journey. Teachers selected their action research focus based on the analysis of their baseline video data from year one.

For example, a teacher observed in her baseline video that her strategies were presented as better than those shared by her students, and she selected the goal of "increasing opportunities for students to share their strategies with their classmates." A teacher who routinely asked students to share their solutions with their peers, but who observed that the student strategies were lacking in variety, selected "anticipating and promoting multiple solution paths for problems." The following is an excerpt from a teacher's learning log. In it, she reflects about why she chose her specific action research goal:

> My classroom is organized chaos. In an effort to conserve time, I tend to show them [the students] my methods as alternatives rather than encouraging them to find more ways themselves. . . . My action research goal is to increase opportunities for students to share strategies with their classmates. They do some of that by working in groups, but I want to see them naturally look for multiple methods of solutions and see how they are connected.

During monthly P^3 sessions, teachers would share a portion of their classroom video with their peers as part of a professional problems-of-practice segment. During this time, the teacher would ask his or her cohort members to watch the video and help in solving a particular problem that she or he had identified. The use of a video-viewing protocol was intended to prompt greater participation of the group members, and it usually resulted in more substantive conversations about the problem of practice. Teachers were also routinely asked to reflect in writing about their videos and progress towards their selected action research goal.

Shift Happens

While the major emphasis of P^3 was the transformation of pedagogical practices, ultimately the ability to develop the kind of problem-based classroom we hoped teachers would achieve required us to engage cohort members in solving rich, challenging problem-based tasks. During interviews with district participants, the enthusiasm for these tasks was expressed by many of the teachers and administrators in the cohort.

Over time, teachers began to share their increased frustration with the textbooks they were using, which they said were "impoverished" when it came to the kinds of rich and intriguing tasks they were experiencing at P^3. "I struggle with having my students develop an intellectual curiosity in doing the math," commented a P^3 teacher. He added that he felt this was directly related to the lack of rich problems in his high school textbook. Another P^3 teacher said, "When my kids see me bring up the P^3 website in class, they get so excited because they know we are getting ready to do a challenging and engaging task!" This shows that the desire to solve cognitively demanding tasks was not limited to the teachers alone, but that students responded to these tasks in very enthusiastic ways as well! These sentiments foreshadowed a major shift that would take place in what was supposed to be the last formal year of the project.

Teachers and Administrators Reflect on *P³*

End-of-project interviews of P^3 participants provided project leaders and DMC members with additional insights into how the work of the P^3 community inspired a vision for a "new normal" in high school mathematics instruction. The research suggested that the inclusion of high school administrators did indeed result in increased attention to and dialogue about mathematics instruction in the participating schools. As one participant put it:

> I think that it is more than just a set of data that is an indicator of progress—rather it is the culture of the classrooms that are "P-cube-like" that has the biggest impact. Our district and my school in particular has embraced the work P-cubed has done with the eight teachers, math specialist, and four administrators involved, and it is now spilling over into our district's math professional development, department meetings, and day-to-day interactions and collaboration.

In addition, P^3 appeared to play a role in promoting conditions for change by "reducing the risks and raising the incentives for teachers to change their practices." One administrator remarked:

> I think they [teachers] are unsure to try something new unless they know that we're going to back them up. . . . We're here, so they know that we understand. So we're not coming in dinging them for maybe not doing it the way they've always done it, [the way they've] always been evaluated or expected to do it. . . . We are now able to kind of be on the same plane with them and give them some more targeted feedback.

Administrators expressed their appreciation for the opportunity for substantive, ongoing engagement in P^3 activities with their teachers. Inviting administrators to be meaningful participants made them feel valued, deepened their commitment, and enhanced their relationships with their teachers. They felt "doing math together" supported complementary shifts toward investigative habits of mind in individuals' reflective practices, in what teachers and students did in their classrooms, in the instructional partnerships of administrators and teachers, and in the culture of their math departments. The administrators also described how P^3 had changed their practices. One of the assistant principals acknowledged how the program had helped him "grow personally to be able to ask harder questions or solicit more in-depth reflection and feedback" (Fifield 2010, p. 8). Another administrator added that the conversations with teachers had expanded from content coverage to include attention to students' thinking and process skills. They described P^3 as a model for supporting their work as instructional leaders and change agents.

> Quite honestly, this is the way it should be. We are supposed to be instructional leaders. Many times within the principal position we're not instructional leaders, we're managers. Disciplinary things, whatever else, we manage that. This is the type of thing we need to be involved in and you know, believe me, these days that I get to come up here, I look forward to them. I truly do. Getting to work with the teachers and being able to change things at that level, that's the type of thing we really need to do to move our schools forward. So these days our involvement I think is absolutely essential. And I'm glad that they invite us here and I'm glad that they look at us as good resources to be here. And being here with the teachers is just invaluable for us. If we could do all our professional development along this type of manner we'd probably be in much better shape.

Ultimately, P^3 provided Delaware's high school administrators with their first real long-term opportunity to develop as instructional leaders in the area of mathematics.

Finally, teachers expressed that "being part of a community of other 'math people' complemented and extended . . . [their] personal work of thinking about math and about themselves as learners . . . By doing math together they built shared commitments to improving teaching and learning. The participants spoke of P-Cubed as a community in which they could explore new ideas and approaches, take risks with the support of peers, and reflect on themselves as educators. By supporting teachers and administrators as learners, P-Cubed modeled what the participants' own departments and classrooms could be like as communities devoted to learning through problem-based mathematics" (Fifield 2010, p. 5).

Moving Beyond P^3

During the three years that teachers and administrators were part of the P^3 learning community, they not only appeared to develop a shared vision of problem-based learning and a better understanding of the pedagogical "habits of mind" for creating a highly effective problem-based high school mathematics classroom, but participants also appeared to feel empowered to legislate for changes in the system that would support a shift to this "new normal." Their well-articulated and pervasive attitudes about the need for better high school instructional materials, the persistence of poor mathematics achievement (60 percent) of tenth graders on the state assessment, and a high degree of curriculum fragmentation with regard to the use of high school textbooks in coalition districts prompted the DMC to commission its director to form a high school curriculum review subcommittee. The committee of secondary teachers from each county in the state was charged with analyzing and making recommendations for all of the high schools in the state to adopt a narrow set of problem-based textbooks. The recommendation came not as a mandate from the state, but rather as a compelling call for action from the district participants and leaders themselves! A newly funded project, *Beyond P³*, would potentially broaden the impact of the professional development to engage every high school mathematics teacher in the state.

In the final months of the *P³* project, the DMC shared the findings of the high school curriculum review subcommittee with district stakeholders and the community. At the time of this writing, seventeen of Delaware's nineteen public school districts, as well as one charter school district, are poised to participate in the coalition's *Beyond P³* curriculum-based project. The four-year project will immerse teachers and their administrators in lessons from one of two problem-based curriculum materials, the *Core-Plus Mathematics Program* (2nd edition) and the *Interactive Mathematics Program* (2nd edition). Project leaders will support deeper content and pedagogical knowledge, and, ultimately, students in nearly all of Delaware's high school classrooms will have access to rich, engaging, and challenging problems to solve. As DMC districts revitalize their approaches to mathematics teaching and learning at the high school level, Delaware may very well be the "first state" in the nation to exemplify the kind of reform-based practices called for in NCTM's *Principles and Standards for School Mathematics*.

......

This chapter is dedicated to the teachers and administrators of the Powerful Pedagogical Practices (*P³*) community, who have inspired a new vision for teaching and learning mathematics in Delaware high schools.

REFERENCES

Ball, Deborah L. "With an Eye on the Mathematical Horizon: Dilemmas of Teaching Elementary School Mathematics." *Elementary School Journal* 93 (1993): 373–97.

Black, Paul, Christine Harrison, Clare Lee, Bethan Marshall, and Dylan Wiliam, "Working Inside the Black Box: Assessment for Learning in the Classroom." *Phi Delta Kappan* 86, no. 1 (September 2004): 8–21.

Boaler, Jo. *What's Math Got To Do With It? Helping Children Learn to Love Their Least Favorite Subject*. New York: Penguin, 2008.

Cohen, Elizabeth G., Rachel A. Lotan, Beth A. Scarloss, and Adele R. Arellano. "Complex Instruction: Equity in Cooperative Learning Classrooms." *Theory into Practice* 38, no. 2 (1999): 80–86.

Fifield, Steve. "Teachers and Principals Reflect on 'P-Cubed' Professional Development: A Qualitative Analysis of Strengths, Effects, and Challenges." University of Delaware: Delaware Education Research and Development Center (November 2010): 1–16.

Grossman, Pamela, Samuel Wineburg, and Stephen Woolworth. "Toward a Theory of Teacher Community." *Teachers College Record* 103 (2001): 942–1012.

Herbel-Eisenmann, Beth A., and M. Lynn Breyfogle. "Questioning Our Patterns of Questioning." *Mathematics Teaching in the Middle School* 10, no. 9 (May 2005): 484–89.

Hiebert, James, Thomas P. Carpenter, Elizabeth Fennema et al. *Making Sense: Teaching and Learning Mathematics with Understanding*. Portsmouth, N.H.: Heinemann, 1997.

National Council of Teachers of Mathematics (NCTM). *Principles and Standards for School Mathematics*. Reston, Va.: NCTM, 2000.

National Research Council. *Adding It Up: Helping Children Learn Mathematics*. Washington, D.C.: National Academy Press, 2001.

National Research Council, Mathematics Sciences Education Board. *Everybody Counts: A Report to the Nation on the Future of Mathematics Education*. Washington, D.C.: National Academy Press, 1989.

Piaget, Jean. "Piaget's Theory." In *Carmichael's Manual of Child Psychology*, edited by Paul Henry Mussen. New York: Wiley, 1970.

Roh, Kyeong Ha. *Problem-based Learning in Mathematics*. Columbus, Ohio: Clearinghouse for Science, Mathematics, and Environmental Education, April 2003. ERIC Document Reproduction No. EDO-SE-03-07.

Schulman, Lee S., and Judith H. Schulman. "How and What Teachers Learn: A Shifting Perspective." *Journal of Curriculum Studies* 36, no. 2 (March-April 2004): 257–71.

Stigler, James W., and James Hiebert. *The Teaching Gap: Best Ideas from the World's Teachers for Improving Education in the Classroom*. New York: Free Press, 1999.

Wilson, Suzanne M., and Jennifer Berne. "Teacher Learning and the Acquisition of Professional Knowledge: An Examination of Research on Contemporary Professional Development." *Review of Research in Education* 24 (1999): 173–209.

Part VII

Sustaining Professional Collaborations

What must be in place to sustain the energy and gains within a professional collaboration?

What sustains a professional collaboration? What factors contribute to the sustenance of a professional collaboration? Why is it that some professional collaborations thrive? The "good work" that comes from professional collaborations can come to an end when individuals in the collaborations change or when the focus of the professional collaboration comes to an end.

As you read the chapters in this part, think about those professional collaborations that you've been a part of and what sustained them. The first chapter here (Groth) provides illuminations into the factors that may or may not sustain a professional collaboration, while the second chapter (Moss et al.) recounts the way that a professional collaboration shifted and changed in order to be sustained.

Use the following reflective questions as you consider the sustainability of professional collaborations:

- What is the balance among individual interests and organizational interests needed to sustain a professional collaboration?

- What happens when individuals change in professional collaborations?

- Compare the factors for sustainability identified in both chapters. In what ways are the factors different, and in what ways similar? In what way do these factors resonate with the experiences you've had in professional communities?

CHAPTER 21

Establishing and Maintaining Collaboration

Factors Related to the Sustainability of Lesson Study

Randall Groth

D URING THE PAST DECADE, LESSON STUDY HAS GAINED MOMENTUM as a means of facilitating collaboration among mathematics teachers. Many versions of lesson study exist, but at its core it involves assembling a group of teachers to plan a lesson, identifying a teacher from the group to carry it out, having the group observe the lesson, and holding a debriefing session where teachers share their thoughts on the lesson's effectiveness (Yoshida 2008). The power of lesson study lies in its potential to help teachers attend to students' mathematical thinking processes during instruction (Stigler and Hiebert 1999).

Although lesson study is a promising means of professional development, it is likely to be more successful in some contexts than others. This chapter describes factors related to the sustainability of lesson study groups. Through considering these factors, those interested in lesson study can anticipate some of the potential inroads and obstacles associated with its implementation.

A Model for Sustaining Lesson Study

The sustainability factors discussed here are drawn from a lesson study project that brought university faculty together with middle and high school teachers from three different school districts. Each school district established at least one lesson study group. The general working model for each group is shown in figure 21.1 and summarized below. I was the university project leader, and this chapter represents my perspective from working alongside teachers and university faculty over the course of two academic years.

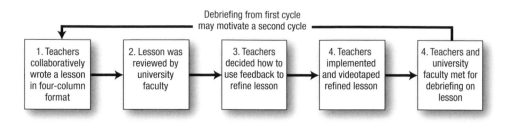

Fig. 21.1. Working structure for lesson study groups in project

To begin a lesson study cycle, the teachers in each group collaboratively wrote a lesson addressing the curricular objectives of their choice. Each of the groups targeted concepts for which their students had shown need for improvement on formal tests. Each written lesson was sent to university mathematics and mathematics education faculty for review. University faculty offered written comments on aspects of the lessons such as problem selection, clarity, and responsiveness to students' thinking. The comments were sent back to each group within two weeks of the initial lesson submission. Each group was then asked to take the comments into account in producing a polished lesson. Teachers had the autonomy to accept or reject any of the advice that was offered. When the final written lesson was finished, a teacher from the group implemented it in his or her classroom. The lesson was videotaped so the entire group of teachers—and university faculty members—could view and discuss it during a debriefing session. These debriefing session conversations generally helped launch to another cycle of lesson study.

Implementation of the Lesson Study Model

The lesson study model described above had varying degrees of longevity in each of the school districts involved. District A participated in two years of the project and sought to be involved for a third year. District B participated in both years but then declined to participate in a third year. District C participated in the first year and was not involved in the second. Because of the varying longevity in each school district, the experiences within each district shed light upon factors that facilitate and impede lesson study.

During year 1 of the project, four lesson study groups were formed: two in District A, one in District B, and one in District C. Information about the groups is summarized in table 21.1. Each group went through two cycles of lesson study (fig. 21.1) during the first year. Year 2 of the project involved one cycle of lesson study rather than two. After completing a cycle of lesson study during year 2, the groups shared their written lessons and reflections in an online forum and set goals for continuing professional development.

Table 21.1

Summary of information about participating lesson study groups

	Year 1 of Project			
	Teacher recruitment	Lesson topic	Grade level focus	Number of cycles done
District A, group 1	Voluntary participation	Algebra tiles and polynomial operations	High school	2
District A, group 2	Voluntary participation	Systems of equations	High school	2
District B group	Voluntary participation	Functions and data collection devices	High school	2
District C group	Mandatory participation	Volume and Pythagorean theorem	High school	2
	Year 2 of Project			
	Teacher recruitment	Lesson topic	Grade level focus	Number of cycles done
District A, group 1	Voluntary participation	Solving equations	Middle school	1
District A, group 2	Voluntary participation	Functions and voltage	High school	1
District B group	Voluntary participation	Matching equations to graphs	High school	1
District C group	Did not participate	Did not participate	Did not participate	Did not participate

Identifying Factors Influencing Lesson Study Sustainability

To document and study the project so that others could benefit from understanding its history, I collected several artifacts:

- Lesson plans written by each group
- Video recordings and transcripts of implemented lessons
- Lesson reviews written by university faculty
- Audio recordings and transcripts of debriefing sessions
- Personal field notes from project planning meetings and debriefing sessions
- Email messages sent to me by participating teachers and administrators
- Anonymous comments from teachers about the strengths and weaknesses of the project, collected via a questionnaire

I became familiar with each artifact because I gathered lessons from teachers as they were produced, organized and attended debriefing sessions, and served as the main point of contact between teachers and the university. As I analyzed the artifacts, a vexing question emerged: What factors contribute to the sustainability of lesson study? I formed my own conjectures about the question and tested them against the data during and after the project.

I used two methods of comparative analysis identified by Corbin and Strauss (2008) to examine artifacts. One method was constant comparison, which involved looking for similarities and differences among key incidents that occurred during the project. Incidents that revealed contrasts among lesson study groups were of special interest because of their potential to explain why lesson study lasted longer in some settings than in others. The second method of analysis was theoretical comparison. Theoretical comparisons involve turning to previous experiences and the literature to make sense of an incident. When incidents strongly resembled those occurring in successful teacher education projects, I inferred that they were important factors in supporting the lesson study groups. Likewise, incidents resembling those associated with less successful professional development described in the literature were identified as factors contributing to the decline of lesson study.

The incidents documented in these artifacts were also examined with their institutional contexts in mind. Cobb et al. (2003) observed that teachers' institutional settings can both afford and constrain reform initiatives. Hence, to identify factors related to the sustainability of lesson study, the scope of analysis needed to extend beyond just the dynamics of the teacher groups involved in the project. It also considered the characteristics of the schools and districts in which the groups were situated. Additionally, the groups had substantial interaction with university faculty in the process of writing and reflecting on lessons. These interactions suggested another layer of analysis of how the university community might afford or constrain lesson study. Hence, interactions within different relevant communities were examined in order to better understand the reasons for the varying degrees of longevity of lesson study in each school district.

Factors Influencing Lesson Study Sustainability

My artifact analysis suggested five primary factors related to the sustainability of lesson study: methods of recruiting teachers, actions and preparation of the lead teacher in each group, relationships between teachers and university faculty, competing initiatives for reform, and participants' understanding of the lesson study model. Each factor is discussed in detail below.

Factor 1: Method of Recruitment

One of the sharpest differences among the lesson study groups was the method by which teachers were recruited. At the outset of the project, teachers from all of the school districts were invited to an introductory session on the purpose and process of lesson study. Because the session was held during the school day, sufficient grant funds were made available for substitute teachers. District A allowed all of its eligible mathematics teachers to attend. After attending the session, District A teachers were to decide if they wanted to join a lesson study group. Participation was not mandatory. However, by the end of the introductory session, District A teachers voluntarily gathered themselves into two lesson study groups and set goals for each group. None of the District A teachers opted out.

Districts B and C approached recruitment differently. Neither district allowed all of its teachers the time to attend the introductory session on lesson study. Only the individuals who were to serve as the lead teachers attended. This arrangement prevented teachers in Districts B and C from gathering into groups and setting goals, as occurred with District A teachers. District B teachers, like those in District A, were given the option of whether or not to participate in the project. All eligible District B teachers ultimately did decide to participate. Teachers in District C, on the other hand, were not given a choice. School administrators mandated participation. At the introductory meeting for the project, the individual designated as lead teacher for District C speculated that teachers at his school would not participate unless required to do so.

The different methods of recruitment to the lesson study project stand out as particularly significant in light of previous literature on the role of teacher autonomy in professional development activities. Castle and Aichele (1994), for example, argued,

> Professional development takes many forms, but true professional development, in the sense of resulting in meaningful and long-lasting qualitative change in a teacher's thinking and approaches to educating, is an autonomous activity chosen by a teacher in search of better ways of knowing and teaching mathematics. . . . Externally imposed professional development activities, though well-intentioned, are doomed to failure. (p. 3)

The self-organization of teachers into lesson study groups in District A stands in stark contrast to the mandated participation in District C. It also contrasts, perhaps on a lesser scale, with District B, where teachers voluntarily participated but did not have the opportunity to cohere around a common goal at the initial project session. Voluntarily cohering around a common goal permits teachers to take a degree of ownership of the project, and identifying a desired goal related to teaching and learning mathematics provides a starting point for autonomous activity.

Factor 2: The Lead Teacher's Actions and Preparation

The lead teachers of each lesson study group performed several different functions. They arranged lesson planning sessions for their groups. They also served as contacts with the university, sending written drafts of lessons to me and helping to schedule debriefing sessions.

The actions of the lead teacher emerged as an important factor in sustaining lesson study because of differences in the way each lead teacher approached his or her role during the first year of the project. In District A, most of the contact between the university and both of the district's lesson study groups occurred through Joan (a pseudonym, as are the rest of the names in this article). She sent detailed plans for both groups in accord with the timeline agreed upon with the university. She found various opportunities to engage her group in planning and revising lessons, such as working lunches and afterschool sessions. Joan became a strong advocate for the lesson study model, recommending that it be extended to middle school teachers during the second year of the project. During the second year, Joan served in the same capacity as the first, leading a high school lesson study group.

District B's lead teacher, John, also served as a point of contact with the university. He fulfilled the obligations of helping to schedule debriefing session times and sending written lesson plans. The level of detail provided in his group's written lessons, however, was substantially less than those from District A. District A provided student handouts and other related materials with their written lessons, but District B provided just a brief outline of the activities to be done during a lesson. University faculty reviewing District B's first lesson remarked that it was difficult to provide constructive feedback given the lack of detail. When a District B administrator read the university faculty feedback, she convened the lesson study group and pressed them to provide more information about what was to be done in the lesson. During the second year of the project, District B appointed a new lead teacher for its group, creating less continuity in the position than District A had.

In District C, there was little evidence that the lead teacher facilitated meaningful conversations during lesson planning. The task of producing the written plans was delegated to the teachers implementing them. As a result, the teachers in the group were not familiar with each lesson until seeing it on video during debriefing sessions. Debriefing sessions became critiques of an individual teacher's implementation of a lesson rather than reflection on areas of improvement that the group could pursue collectively. James, the lead teacher for the group, placed low priority on attending the debriefing sessions, scheduling other afterschool meetings that overlapped with the times the sessions were scheduled. At the end of the first year of the project, James acknowledged the group's low fidelity toward proper implementation of the lesson study model.

The variation in effectiveness among the lead teachers suggests that additional professional development for the lead teachers may have been helpful. For instance, Joan, John, and James could have conversed with one another to compare their experiences. John and James both expressed the desire to be effective facilitators. Joan had useful strategies for lesson study facilitation, so sharing those ideas with John and James may have been helpful. Although this sort of communication would place additional demands on the lead teachers, time dedicated to becoming an effective lead teacher is an investment in the life of one's lesson study group. Effective lead teachers can prompt their groups to examine their personal content knowledge, assumptions about students' mathematical thinking, and goals for student learning (Lewis, Perry, and Hurd 2009).

Factor 3: Relationships with University Faculty

The lesson study model used for the project provided many opportunities for interaction between teachers and university faculty. Interactions occurred during reviews of written lessons and debriefing sessions. These interaction opportunities were built into the structure of the project because past research illustrated that outside perspectives can prompt a lesson study group to consider fruitful ideas it may not on its own (Fernandez 2005). The nature of the interactions between university faculty and lesson study groups varied from district to district.

Several teachers in District A had previously worked with the university faculty involved in the project. Some of them had taken graduate courses and workshops taught by the university faculty. During the lesson study project, lesson reviews and debriefing sessions tended to elicit diverse points of view about pedagogy and content in District A. Joan's group, for example, focused on using algebra tile manipulatives to represent polynomial factoring and multiplication geometrically. During their first lesson, the group planned to have students check the accuracy of their algebra tile solutions by using previously learned methods for factoring and multiplication. In writing reviews and participating in debriefing sessions, university faculty argued that such use of the algebra tile model did not exploit all of its geometric characteristics. In particular, they encouraged teachers to revisit factoring and multiplication with integers first (e.g., 3×5 represented with unit blocks) in order to build a foundation for representing more abstract quantities, such as $(2x + 1)(x + 4)$, using algebra tiles. After a great deal of discussion between the lesson study group and university faculty about the relative merits of this approach, teachers used some of the advice in revising the lesson and teaching it a second time. The second lesson began with a review of visual representations for multiplication and factoring situations involving integers, before moving on to problems with variables.

In District B, fewer instances of meaningful conversation between university faculty and teachers occurred. Teachers in the lesson study group listened to the perspectives on content and pedagogy offered by the faculty, but they did not engage in extended discussion about alternative approaches, as in District A. Debriefing sessions consisted of exchanges of ideas, but generally lacked the element of one individual's idea building on another's. Part of the reason for the differences in interaction is likely that no previous professional relationships existed between university faculty and teachers in Districts B and C before the lesson study project began. Teachers in Districts B and C dutifully attended lesson study activities and debriefing sessions and participated to the extent requested, but they were really just in the beginning stages of forming their relationships with university faculty. Trust had yet to be firmly established between university faculty and the teachers at the outset of the project.

The willingness to engage in conversations with outside observers such as university faculty stands out as a particularly salient factor in light of previous research on lesson study. Fernandez (2005), for example, found that an outside observer who was knowledgeable about mathematics helped to prompt teachers to study the content more deeply. While participating in a debriefing session, the outside observer noted that the teacher conducting the lesson was confused about its mathematical content. The teacher took the outside observer's comments back to a meeting with the lesson study group and subsequently developed deeper mathematical understanding of the content of the lesson. Yoshida (2008) agreed that outside observers, or "knowledgeable others," can help teachers develop mathematical understanding during lesson study, and also noted that they can introduce teachers to new curriculum materials and pedagogical techniques. Knowledgeable others can inject life into a lesson study group by sharing ideas the group may not consider on its own, provided the teachers in the group are open to the active exchange and debate of ideas. These types of productive conversations seem more likely to occur if the outside observers have existing professional relationships with lesson study group teachers and have established the element of trust, as was the case in District A.

Factor 4: Competing Initiatives for Reform

Another factor contributing to the lower levels of engagement in lesson study in Districts B and C was the presence of many other initiatives for reform in their schools. An administrator from District B cited a new reform initiative as the primary reason for not seeking to participate in a third year of lesson study. The District B initiative was based on constructing individual professional development plans for teachers rather than having them work together in groups.

District C understandably viewed the project cautiously from the outset because they had just finished participation in a grant with another university for which they were assigned to be a control group. As a control group, they received little benefit from the project or contact with university faculty. They expressed the concern that lesson study would be a similar experience. I addressed this concern during initial conversations with the group, ensuring them they would not be used in the same manner for this project. During the lesson study project, some District C teachers were also participating in a professional development project focused on the use of technology in teaching mathematics. One of the activities for the technology professional development was to write a lesson and videotape it. District C teachers used one of the lessons developed and videotaped for the technology professional development as the second lesson they submitted for the lesson study project. The decision to submit a prewritten and already video-recorded lesson meant there was no opportunity for university faculty to provide feedback on the lesson prior to its implementation. The District C teachers explained their decision by voicing the need to "kill two birds with one stone" in order to satisfy all of their professional development requirements.

District A was not without other professional development projects, but teachers had more autonomy to opt out of them. Before beginning the lesson study project, they opted out of a professional development project with another university, one which involved extensive videotaping and reflecting upon individual lessons. After participating in lesson study, they remarked that they found it to be more effective for team-building than the professional development model focused on individual reflection.

In sum, care must be taken so that lesson study does not become just another item on teachers' schedules. Individuals in districts that already have a number of professional development projects need to examine the scope of projects in which teachers are involved. Efforts to reform schools do not occur in a vacuum. Tyack and Cuban (1995) noted, "If the aims of reforms seem vague, contradictory, or unattainable, educators often respond by turning reforms into something they have already learned how to do" (p. 64). If lesson study is to be worthwhile, difficult choices may have to be made about eliminating obligations to existing incompatible projects. If lesson study is simply added as another layer to a professional development plan, it may not survive. At the very least, the integrity of the lesson study model may be compromised as teachers attempt to make it fit with existing incompatible initiatives.

Factor 5: Participants' Understanding of the Lesson Study Model

When viewed within the landscape of other existing professional development models, lesson study can be difficult to understand. One source of confusion comes from how lesson study emphasizes collective reflection about a lesson to a greater extent than individual reflection. Administrators in all three school districts expressed concern about whether each individual teacher would implement the lessons written by their groups. Although individual implementation of a lesson can be an outcome of lesson study, it is not the ultimate goal. The primary focus is upon group pedagogical problem solving, and it is expected that the problem-solving skills teachers develop will transfer to designing other lessons, even if they involve different content. In some cases, it may not be feasible for each teacher to implement the exact group plan because of differing teaching assignments, but such a situation does not automatically block the implementation of lesson study.

Lesson study also stands in contrast to forms of professional development that aim to have teachers produce a number of lessons in a relatively short amount of time. Even the administration from District A, having recently participated in a project emphasizing quantity of lessons, began with the expectation that entire units of lessons would be produced. They relinquished this expectation, however, after teachers reported seeing value in the in-depth discussions that took place over a single lesson. Lesson study groups can ultimately produce many volumes of lessons based on their experiences working together (Stigler and Hiebert 1999), but at the outset of a project it is important to emphasize the quality of collective reflection over the quantity of lessons produced.

Another feature of lesson study that stands in contrast to other forms of professional development is the role of group autonomy. In the project structure described above, university faculty served as advisors but did not dictate the content and pedagogy of the lessons produced. This arrangement was unsettling to the District C lesson study group, and at the outset they asked university faculty to specify the content they should focus upon. Some of the District C teachers also believed at the outset that the lead teacher would be taking a course at the university and then reteaching the information obtained to the group. The notion that lesson study begins with the ideas of the group, rather than those of an outside expert, is a fundamental mindset that must be present if lesson study is to flourish (Stepanek 2003).

Because lesson study is more complex in comparison to many other forms of professional development, it should not be taken for granted that all involved understand the process and its purposes. It is important for teachers to understand that the process is structured to help turn classrooms into research sites and to foster deep professional conversations rather than just to improve lesson-plan writing skills (Tolle 2010). If vital elements such as collective reflection are neglected or underdeveloped, lesson study loses much of its power. The phrase "lesson study" sounds deceptively simple on the surface, so it is important to help teachers look deeper to understand how neglecting steps in the process can lead to its downfall. The cases of Districts A, B, and C illustrate the importance of understanding and executing each step.

Bringing the Five Factors Together: Advice for Establishing Sustainable Lesson Study Groups

Taken together, the five factors distinguishing the three school districts from one another suggest a potential action plan for establishing sustainable lesson study groups. The components of the plan and how they relate to one another are illustrated in figure 21.2. District A modeled each component shown in the figure. Districts B and C modeled some components but lacked others, such as voluntary participation, high priority on the project, and continuous open and critical conversations. The absence of any of the components, such as a supportive administration, an effective lead teacher, and relationships with trusted knowledgeable others, may shorten the life of a lesson study project even if other components are in place. Additional details about each component and a suggested sequence of implementation are provided below.

A good starting point for establishing a lesson study group is to identify an effective lead teacher and trusted outside consultants. A lead teacher's actions can have a substantial impact on the longevity of the group. Communication skills and commitment to the process are essential. The longest-lasting groups in the project described above were those in which the lead teacher actively fostered communication among group members and also communicated effectively with university faculty. The lead teacher may also have pre-existing relationships with university faculty, or mathematics teachers at other schools, who can be drawn upon as "knowledgeable others." Knowledgeable others who have expertise in mathematics or mathematics education can contribute perspectives that may not otherwise be present among lesson study group members.

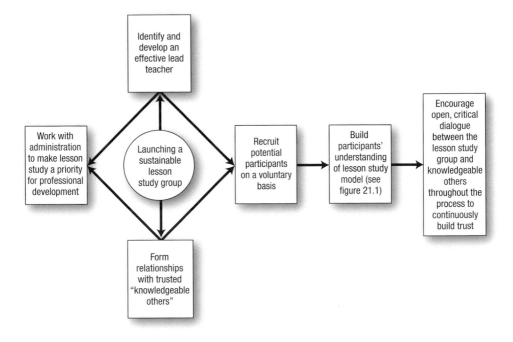

Fig. 21.2. Flowchart of a plan for establishing sustainable lesson study groups

Once a lead teacher and knowledgeable others have been identified, initial work in establishing the lesson study group can progress along two fronts. The first front involves communicating with administrators to ensure that lesson study has a prominent place among the professional development initiatives in place in the school district. If teachers are overwhelmed with a variety of other initiatives, and are not able to opt out of any of them, lesson study may be given a very low priority, as in District C. If at all possible, participation in lesson study should be voluntary, since mandating it as a form of professional development can cause distrust and disengagement. As the groundwork for lesson study is laid with the administration in a school district, work along the second front can begin: recruitment of teachers. Recruiting a core of teachers who have pre-existing relationships with knowledgeable others can greatly help catalyze lesson study, as in the case of District A.

An important aspect of recruitment is helping teachers understand the nature of the lesson study model. Figure 21.1 can be used as a starting point in this endeavor. Participants should understand that group members, lead teachers, and knowledgeable others all work together in directing the group. Ideally, all potential participants should be given the opportunity to participate in an initial meeting. At the meeting, the general structure of the lesson study model to be used can be shared, and teachers can be given the opportunity to identify possible instructional goals for their work together as a group. For the teachers in District A, cohering around common instructional goals at an initial meeting provided a good starting point for sustaining their work with one another. Teachers in Districts B and C did not have the opportunity to do so.

After lesson study groups have been established, open and critical dialogue should be constantly encouraged between the teachers in the lesson study group and the knowledgeable others involved in the process. Each group of individuals has a unique perspective to contribute that should be elicited and valued. Teachers have knowledge of the specific school context in which the lessons are taught, and outside observers who are knowledgeable about mathematics and mathematics education can broaden the pedagogical perspective of the group and prompt deeper understanding of content when necessary. As lesson study groups engage in vigorous discussion of optimal strategies for teaching mathematics, they can become self-sustaining communities of pedagogical problem solving and intellectual renewal for participants.

REFERENCES

Castle, Kathryn, and Douglas B. Aichele. "Professional Development and Teacher Autonomy." In *Professional Development for Teachers of Mathematics*, 1994 Yearbook of the National Council of Teachers of Mathematics (NCTM), edited by Douglas B. Aichele, pp. 1–8. Reston, Va.: NCTM, 1994.

Cobb, Paul, Kay McClain, Teruni de Silva Lamberg, and Chrystal Dean. "Situating Teachers' Instructional Practices in the Institutional Setting of the School and District." *Educational Researcher* 32, no. 6 (2003): 13–24.

Corbin, Juliet, and Anselm Strauss. *Basics of Qualitative Research,* 3rd ed. Thousand Oaks, Calif.: Sage Publications, 2008.

Fernandez, Clea. "Lesson Study: A Means for Elementary Teachers to Develop the Knowledge of Mathematics Needed for Reform-Minded Teaching?" *Mathematical Thinking and Learning* 7, no. 4 (2005): 265–89.

Lewis, Catherine C., Rebecca R. Perry, and Jacqueline Hurd. "Improving Mathematics Instruction through Lesson Study: A Theoretical Model and a North American Case." *Journal of Mathematics Teacher Education* 12, no. 4 (2009): 285–304.

Stepanek, Jennifer. "Researchers in Every Classroom." *Northwest Teacher* 4, no. 3 (2003): 2–5.

Stigler, James W., and James Hiebert. *The Teaching Gap: Best Ideas from the World's Teachers for Improving Education in the Classroom.* New York: Free Press, 1999.

Tolle, Penelope P. "Lesson Study: Still a Work in Progress in America." *Mathematics Teacher* 104, no. 3 (2010): 181–85.

Tyack, David, and Larry Cuban. *Tinkering toward Utopia: A Century of Public School Reform.* Cambridge, Mass.: Harvard University Press, 1995.

Yoshida, Makoto. "Exploring Ideas for a Mathematics Teacher Educator's Contribution to Lesson Study." In *The International Handbook of Mathematics Teacher Education*, vol. 2: *Tools and Processes in Mathematics Teacher Education*, edited by Dina Tirosh and Terry Wood, pp. 85–106. Rotterdam, the Netherlands: Sense Publishers, 2008.

CHAPTER 22

Building and Sustaining Professional Collaborations

Using Japanese Lesson Study to Improve the Teaching and Learning of Mathematics

Joan Moss
Richard Messina
Elizabeth Morley
Diane Tepylo

THIS CHAPTER HAS A DOUBLE FOCUS. One is the chronicling of the evolution of a professional learning community engaged in lesson study. This story includes the challenges the group faced, its recommitment to the process despite these challenges, and the innovations the group forged that sustained it over the next six years. The second focus is an account of the design process of a single, representative lesson developed by the members of this learning community. The process and results of the design and implementation of this grade 2 lesson, which concerned the addition of positive and negative numbers on a number line, illustrates how the group's innovations in the lesson study process were successful both in fostering students' understanding of integers and in giving the teachers important insights into mathematics for teaching and thus sustaining their ongoing interest in the lesson study approach. This lesson study case addresses some of the issues central to professional learning communities. These issues include facing challenges in professional development and related opportunities for learning; gaining and nurturing teachers' ownership of, and engagement in, collaborations; and employing structures that can support and sustain teachers' active participation over time.

We, the authors, are, respectively, a mathematics education researcher, a classroom teacher, an elementary school principal, and a doctoral student in math education. We are all affiliated with the Ontario Institute for Studies in Education at the University of Toronto, and we have all, in different roles, been involved with this lesson study project. We begin our chapter with a brief discussion of Japanese Lesson Study in North America, and the challenges it has faced in this setting. Then we document how the staff in a university elementary laboratory school took up the lesson study process. We highlight three significant adaptations the group forged that appear to have contributed to the sustainability of the project. Finally, we discuss the design, implementation, and results of a representative lesson designed by members of the lesson study group.

Lesson Study and Professional Collaboration

Recent years have seen a dramatic increase in North American settings in the practice of lesson study (Perry and Lewis 2009), a practice that involves collaborative planning, teaching, and reflecting on classroom lessons. The cycle of the four steps that constitute lesson study are: (1) goal setting/investigation; (2) planning; (3) implementation/research lesson; and (4) debriefing/reflection (Lewis, Perry, and Murata 2006). Credited with substantial gains in mathematics learning in Japan (e.g., Fernandez, Cannon, and Chokshi [2003]; Fernandez and Cannon [2005]; Lewis [2002a]), lesson study has gained the attention of researchers and educators as a powerful professional development model because of its classroom context, its ongoing collaborative process, its focus on students, and its attention to teachers' concerns and questions (Bruce and Ladky 2010). Additionally, Stigler and Hiebert (1999), in describing the benefits/potential of lesson study, assert that it provides a unique way for teachers to look at their own practice with new eyes.

Research has found that, despite the known benefits of lesson study and educators' increasing familiarity with the process, there appear to be considerable obstacles to the success of lesson study in North America (e.g., Fernandez and Cannon [2005]; Lewis [2002a]). One obstacle is the sustainability of lesson study over time. In particular, Murata has noted that in most cultures outside of Japan (such as in the United States), professional development is usually experienced as discrete and separate programs implemented from one year to another. In contrast, lesson study in Japan operates as a sustained effort that continues over years and becomes rooted in the teaching culture as an integral process in teachers' professional lives (Murata, in press). Although the challenge of sustainability besets many professional development initiatives, the particular challenge for lesson study in North America is that it requires a shift in mindset from discrete, year-to-year programs to an ongoing process.

Experimenting with Lesson Study: Troubled Beginnings

The schoolwide lesson study project began in fall of 2004 with the teaching staff in a university laboratory school in downtown Toronto. The school is composed of nine classes of twenty-two students each, from preschool to grade 6, with each class representing a wide range of abilities. Each class has one teacher and, for part of the year, an intern student teacher. The school also has specialty teachers including art, French, physical education, and special education. Having heard reports of the early North American adaptations of lesson study, in particular from the initiatives of the Mills College Lesson Study Group (e.g., Lewis [2002b]; Lewis and Tsuchida [1998]) and the work of the Fernandez and colleagues group at Teachers' College at Columbia University (e.g., Fernandez and Yoshida [2004]), several teachers and the principal proposed lesson study for the laboratory school.

A number of features of lesson study appealed to the teaching community at the school, particularly the collaborative nature of the process and the focus on observations of children's reasoning. However, once the school community began to engage with the lesson study process, what became immediately apparent was the tension that emerged between the teacher's customary discourse practices—typically focused on general pedagogy—and the discourse in lesson study, which is only intended to address the specific aspects of specific lesson goals and design elements.

Furthermore, as the teachers moved towards the planning stages, many became skeptical about some of the major components of lesson study—a situation that led to significant alterations to some of its fundamental principles. One aspect was a matter of philosophy: for many of the teachers in the school, the focus on the design of a *single* lesson seemed to contradict the school's implicit philosophy that the focus should be on a full unit of study. Thus, when the members of the school community first undertook lesson design, rather than following the customary lesson study practice of concentrating on a single lesson, they planned a longer unit of study.

A second concern was that teachers felt they were developing a closed, highly specific lesson script. Consequently, as the group engaged in their lesson planning, the teachers designed the lessons based on more open-ended ideas of how the lesson might be taught rather than on specific trajectories.

A third and very important issue with the process was the lesson study's inclusion of live classroom observations by other teachers and outside guests. The teachers felt this might be too intrusive for the students. To address this concern, it was decided that instead of observers in the classrooms with children, the lessons would be videotaped and reviewed later by the participants and discussant.

These concerns, and the consequent alterations to the lesson study process by the school staff, exemplify Fullan's (1993) caveat that educational terms travel well, but the underlying principles often do not. By the time the teachers had modified the process to suit their levels of comfort and familiarity, they had departed so substantially from the lesson study process as to render it nearly unrecognizable. In the end, this first-year effort was unsuccessful, and the teachers were disheartened. They found the lesson study process time-consuming and burdensome, with little that had been learned by either the teachers or the students.

A Second (Successful) Try at Lesson Study

Although the group still had questions about the need for all of the steps and formalities of the lesson study approach, they still saw merit in the *goals* of lesson study and therefore were willing to learn more about the process. In particular, teachers were interested in how the structures of lesson study could be applied in their own setting and help them, as teachers, to become more knowledgeable about the teaching and learning of mathematics.

Learning about Lesson Study

In the fall of the second year the school invited a well-regarded lesson study researcher/practitioner, Aki Murata, to come to Toronto to lead the school staff in a one-day in-depth introduction to the philosophy and practices of lesson study. As a result of the teachers acquiring a much deeper sense of the potential of the lesson study process, the school staff committed to a new start and agreed to follow the detailed protocol of the full lesson study cycle. At this time they invited a mathematics education researcher (the first author of this chapter) to establish a database to track the development and use of lesson study in their school community. At the time of writing, the school is in the beginning of its sixth year of continuous lesson study—the first pilot year discussed above, followed by five successful years.

Establishing New Lesson Study Routines and Schedules

In that year following Murata's visit the group established a number of protocols/practices and schedules for lesson study practice that continue to the present time. First, it was decided to have two lesson study groups (LSGs) with all of the teaching staff in the school belonging to one or the other. As for the schedule: each year, in the fall and in the spring, each LSG participates in a full lesson study cycle and designs and implements a new research lesson. The school staff set aside two full days per year (fall and spring) when these lessons are publicly presented. These public lesson days follow a specific protocol: in the morning members of the two LSGs distribute lesson plans and observation guides that they create for the visiting teachers and discussants. Then the two research lessons are presented consecutively, with one member of each LSG teaching the specially designed lesson to a class of students while being observed by the rest of the LSG and guests and discussant. The morning lessons are followed by an afternoon colloquium session in which members of the LSG reflect on the lessons, and then other observers, followed by a discussant, draw out implications for lesson study and for teaching learning more broadly. (For more detail on the process, see Moss [2007].)

Researching Our Practice: Design Research Framework

Through this process, the school staff has designed and implemented more than twenty math research lessons covering a wide range of mathematical topics for the full span of age groups in the school. We have also been carrying out an ongoing and extensive data collection encompassing video data from teachers' planning meetings, the public research lessons, and the debrief sessions; artifacts produced by the teachers such as lesson plans, written rationales for the lessons, research gathered on the selected topic, observation checklists, and email correspondence; and student work (collected for selected studies). We have employed the Design Research approach of Kelly and Lesh (2000) and the analytic methods of Collins et al. (2004) to trace the evolution of the lesson study process and to focus on the "products" of this work (Wood and Berry 2003). These "products" include not only materials such as lessons for dissemination, presentations, and publications, but also products of a more theoretical nature, such as the innovations and adaptations to the lesson study process that we describe in this chapter. We chose the Design Research methodology because it allows us to codify and analyze teaching and learning in complex classroom contexts over an extended time. Collins, Joseph, and Bielaczyc (2004) make explicit connections from the Design Research framework to lesson study, describing the iterative processes of testing and refining inherent in both.

At the outset, our research focused on change and growth among the school staff in their level of comfort with and knowledge of mathematics for teaching (Ball et al. 2008). To do this, we examined changes in the teachers' use of mathematical language both in planning and debrief meetings, as well as in the teachers' production of lesson plans and written materials. In the written materials the teachers produced and in the transcripts of the planning and debrief sessions, we found notable shifts from general language to a more specialized use of mathematical vocabulary in the use of mathematical language.

Our research focused on tracking the innovations and adaptations that the group forged over the years. Our goal has been to contribute to Lewis et al.'s (2006) call for more research to identify mechanisms in the lesson study process that may support its sustainability and promote teacher learning. In the remainder of this chapter, we present our findings on the adaptations and provide a detailed discussion of a specific lesson to highlight these adaptations in practice.

Findings: Sustaining Lesson Study through Adaptations

Three adaptations evolved over the five years, and, unlike the adaptations in the initial year, these informed adaptations significantly advanced the function of the group and therefore the sustainability of the lesson study. Given that sustaining change in general (and the lesson study process in particular) is challenging, and given the ambivalence with which the teachers initially began this project, these adaptations that led to sustained lesson study are significant. The three adaptations can be summarized as: (1) broadening the composition of the LSGs; (2) expanding the criteria for selecting mathematics topics; and (3) inserting a practice lesson, an extra step in the usual lesson study process.

First Adaptation: Broadening the Composition of the Lesson Study Groups

As we mentioned above, when the school staff recommenced their lesson study work in the second year, they chose to establish two LSGs that matched the grade levels of the students. In keeping with common lesson study protocol, the two LSGs represented different school divisions. One LSG was composed of primary teachers (pre-kindergarten to second grade) and the other, the junior grade teachers (third to sixth grade). The school specialty teachers (such as art, French, and physical education) also moved into the groups congruent with the ages of the children they were teaching. In year 3, the teachers decided to expand the grade-level span so that each LSG had teachers spanning pre-K to grade 6. One very clear benefit of this new grouping of teachers was the diverse range of knowledge that was brought to their lesson design process.

Second Adaptation: Expanding the Criteria for Selecting Mathematics Topics

With the broadened composition of the LSGs came a shift away from selecting mathematics topics specific to the curriculum, and regarded as challenging, in favor of topics that crossed grade levels and were inspired by shared teacher interests. Initially, the LSGs chose topics that aligned with specific curriculum topics. Over time, however, teachers considered broader topics (e.g., proportional reasoning, understanding mathematical relationships though music, and, as we elaborate below, integers on a number line). This expanded perspective opened up the possibility for the teachers to explore diverse mathematical ideas that they might not have otherwise encountered. This teacher-initiated adaptation enriched the lesson study process for the teachers (and for the students).

Third Adaptation: Insertion of the Practice Lesson

As noted earlier, in their second year of lesson study teachers committed to follow the four-step lesson study process. In step 1, *goal setting/investigation*, the teachers select two topics and align goals for the lesson study. Next, in step 2, *planning*, the LSG examines samples of students' work on the chosen topic, and studies different textbooks' treatments of the topic, which members then use or adopt for their lesson. The third and fourth steps—*conduct research* and *debrief*—involve the LSG in the public teaching and debriefing of the specially designed lessons, with guests and members of the LSG collecting data on students' learning.

As the teachers gained more experience with lesson study, they decided to expand the process by adding an intermediate step between steps 2 and 3. This new step became known as the *practice lessons*—a series of lessons carried out in different grades in advance of the public lesson. Members of the LSG began the practice of experimenting with variations of the research lesson in their own classrooms. These experimental practice lessons gave members of the LSGs the opportunity to determine the appropriate grade level for the final research lesson, to explore the effectiveness of various representations, and to decide on appropriate revisions and refinements. Indeed, our review of our data over the five years revealed that the practice lessons not only led to significant revisions of the final lessons, but also, in the majority of cases, to the LSGs reconsidering the appropriate grade level for the lesson. A particular advantage of the iterative nature of the practice lessons was that it led the teachers to a higher level of analysis of the mathematical topic itself and of its pedagogical implications, and thus a more enriched debriefing session on the public lesson days.

Integers in Grade 2 on a Life-Size Number Line: The Adaptations in Practice

As an example of what these adaptations/innovations looked like in practice, we present the case of the Integer Lesson—one of the two lessons created in the eleventh cycle of lesson study. We summarize each of the phases of lesson study, including the additional practice-lessons process.

Step 1: Goal Setting

At the outset of this lesson study cycle the school staff met to select the topics for the term (fall 2009). In the case of the integer lesson, the selection process began with a suggestion from the physical education teacher to explore the connections between mathematics learning and movement. This idea appealed to others in the group. One teacher mentioned her interest in Howard Gardner's ideas of multiple intelligences, kinesthetic learners, and the implications for mathematics learning. The grade 5 and 6 teacher added that he had heard about research suggesting that gesture may play an important role in mathematics learning. The grade 2 teacher also mentioned a movement game, a form of hopscotch, that she was using with her students as they consolidated addition and subtraction learning. The foregoing were some of the comments that led the group to the decision to explore number relationships through body movement and location in space. The idea of integers came up when the grade 1 teacher recounted a story of a routine classroom calendar activity in which one of her students proposed including a negative number as a creative solution strategy. As the discussion of these ideas went on, the teachers became intrigued with a central question: What could primary students understand about integers, a topic normally introduced to much older students, if mathematics learning were to be integrated with movement? Based on these discussions, an LSG was formed. It included the pre-kindergarten teacher, art teacher, librarian, school principal, several grade-level teachers, and the physical education teacher, and was charged with exploring and designing an introductory lesson on integers and movement.

Step 2: Planning

Once the LSG had decided on movement and integers, the next step was to find resources offering theories and teaching approaches to support students' introduction to integers in grade 2. Generally, at this stage of the design process, the LSGs would consult a variety of math textbooks to find published lessons to draw from. In this case, however, the LSG, though unable to find appropriate resources for grade 2 students, did find lessons for older students that, among other representations, used a thermometer metaphor as an approach to teaching integers. The librarian noted that, because this lesson was to be taught in the fall, the students would be accustomed to the November temperature fluctuations above and below zero Celsius. After considerable discussion, the LSG decided on the context of weather and a thermometer representation because of the real-world connections. A large thermometer was constructed with two sections, red for the positive integers above zero, and blue for negative integers below zero. The general plan for the lesson involved students in moving along the thermometer in directions and increments randomly assigned by the throw of color-coded dice—red for positive and blue for negative.

Step 2.5: Practice Lessons

Practice lesson 1. To test the effectiveness of their lesson plan, the LSG presented a practice lesson in a grade 3 classroom—the targeted grade level for the lesson. As the teachers embarked on teaching this practice lesson, they immediately found it was not working as expected. There were two problems. The first was the context of weather: the students appeared to be more interested in discussing weather fluctuations (and storms they had experienced) than in trying to make sense of the properties or concepts of integers. The second problem was the use of color in both the die and the thermometer: instead of considering directionality on the number line, the students become preoccupied with color as a indicator of movement.

The experience of this practice lesson provided the members of the LSG with a broader, more nuanced understanding of the issues arising from the use of real-world contexts. The issue of real-world applications in mathematics teaching then became a central focus of the LSG's subsequent discussions as they worked on revisions to this lesson. The teachers had assumed that embedding mathematics in real-world contexts would naturally make the mathematics meaningful. Forced by the results of the first practice lesson to reconsider this assumption, the teachers decided to eliminate the thermometer and to try a more abstract context, a number line.

Additional practice lessons. In a subsequent practice lesson the LSG discovered that this new number-line context did indeed support the students in gaining an initial understanding of integers. And, based on further evidence gleaned from a third practice lesson conducted with grade 2 students, they refined the lesson further and added an additional component. In our view, the planning process for the integer lesson was enriched by the diversity of the LSG's interests and the breadth of their collective experiences and mathematical knowledge—all of which were deepened and strengthened by the additional iterations of the practice lessons.

Step 3: Implementation/Research Lesson

The public lesson was taught in second grade by the pre-kindergarten teacher with the help of the physical education teacher. The scope of this paper does not permit an account of the full lesson design, but briefly, the lesson focused on the addition of integers through movement on a floor-sized number line in response to challenges on prompt cards. First in the sequence, a student was given a card (the Start Card) with a number that indicated a starting position on the number line. A second card (the Add/Moving Card) was given next. The student then had to determine the sum and move to the appropriate space on the number line. To begin the lesson, the students sat on the floor around a number line calibrated from 0 to +8 and the teacher led the students in a specific series of challenges. The initial prompts in the game were sequenced so that after two warm-up rounds, where the students were asked only to move to sums of positive integers, a negative integer card was introduced as a Move Card. Specifically, this challenge was to determine and move to a number on the line that corresponded to the sum of positive 2 and negative 4. The conjecture was that, although the students had not been formally introduced to number lines with a negative extension, through the specially sequenced activities, and because of the students' comfort with the operation of addition, the students would naturally see how the 0 to +8 number line would need to be extended to include negative integers. (In anticipation of this, the LSG had already constructed another number line, calibrated from 0 to -8, which they placed out of sight of the students.)

As the transcript below reveals, the students did recognize a need for an extension of the number line:

Teacher:	[*Holding up a card for the children to see*] What is our start number?
Children:	[*In chorus*] Two . . .
Teacher:	[*Holding up a second card and pointing to it*] What is the moving number?
Children:	[*At once*] Minus four, no, negative four, minus four ...
Teacher:	So then what will the end number be? Who will try? [*One of the students, Sam (a pseudonym) stands on the number line on positive 2 as the other children make suggestions.*]
Student 1:	Maybe go back to the starting.
Sam:	[*Moving two steps back to stand on zero; he looks puzzled*] But we don't have enough numbers. We need more numbers. We need more number line. There would have to be minuses.
Student 2:	We need minuses.
Student 3:	[*Pointing to the floor just below the start of the 0–8 number line*] They would go here.
Teacher:	Well ... I just happen to have more number line. [*Teacher places this extension on the floor, and Sam continues to walk backwards and stands at negative 2.*]
Student 5:	Now you can see that the answer is negative two.

Overall, the children appeared to be happy to participate and engaged with the lesson. As a final piece of the lesson, the teachers gathered the students back to the central number line (students had worked in pairs on individual number lines -8 to +8) and posed the following question:

Teacher: If my number was a negative number, any negative number, and I wanted to get to positive six, what could my moving number be?

Many students put up their hand to answer with suggestions; below are the responses of three of the students.

Student 5:   I am imagining it [the negative integer] as minus three, so the answer is nine.

Student 6:   I am imagining it as negative 100, so positive 106.

Student 7:   The negatives go on and on and never stop, just like the positives.

Step 4: Debriefing/Reflection

Both the members of the LSG and the visiting observers agreed on the success of the flow of the lesson, the sequencing of the activities, and generally that the lesson design and particular challenges had supported the students' understanding. As one visiting teacher commented: "One thing that I loved about the lesson is that negative numbers were introduced in such a natural way, almost as a by-product of moving on the number line. It was if you are moving that far in that direction on the number line, there have to be numbers down there."

The discussant for this public lesson, a research mathematician, offered several positive comments on the choice of the number line representation. He noted that this representation not only supported the students' understanding, but also allowed them to communicate this understanding physically, as shown by the way the students naturally oriented themselves on the number line as they responded to the challenges: "I was so intrigued to see them walking forward going up the number line for positive numbers and walking backwards down for negative numbers."

The discussant also noted how the final probes designed by the LSG enabled students to express their understandings and insights, thereby serving to inform the teachers. As a wrap-up at the end of the lesson and as a way to encourage the students to reflect, the teacher (from the pre-kindergarten), asked the (grade 2) students to consider how this same lesson might work for her very young students. The children agreed that the first part of the lesson (adding positive integers walking on a number line) would be fine for her pre-kindergarten students, but the second part of the lesson, the addition of negative numbers, would not work because, in the words of the grade 2 students: "JK children don't know subtraction." This response impressed the discussant, indicating to him that the grade 2 students had made a significant mathematical connection between negative numbers and subtraction.

Reflections on Sustaining Lesson Study

The evolution of the lesson study process and the innovations that were generated and adopted by the teachers themselves—group composition, topic selection, and practice lesson—seems to have afforded the teachers with diverse and frequent authentic learning opportunities that were engaging and served to support the sustainability of this five-year project.

Broadening the composition of the LSGs. When the teachers began that first year of committing to practice the full cycles of the lesson study, they spontaneously moved into two LSGs that represented similar grade levels and school divisions. As time went on they began the practice of sorting themselves into one of two groups based on their individual interests—a process that continues to this day. This new way of grouping themselves appears to have created a cross-fertilization of ideas, experiences, and knowledge; this, in our view, has served to increase the staff's enthusiasm and learning, and to have sustained their engagement in the project. Davis and Simmt (2003, 2006) make compelling arguments grounded in complexity theory regarding the affordances of diverse groupings: in their words, "Each learner has access to a diversity of interpretations and strategies to make sense of the concept or task at hand."

Expanding the criteria for topics. Along with the new mix of LSG membership, and afforded by this new diversity, came the natural broadening of potential subject areas and mathematical topics for the teachers to explore. The effect of this change has been profound. As we look at the evolution of the staff's mathematical discourse, as analyzed through reviewing the videotapes of their discussions in planning and debriefing meetings, it is clear that because of the broad range of topics selected over the years, many of them challenging, and because of the depth at which the LSGs come to know these, there has been significant overall growth in the teachers' Mathematical Knowledge for Teaching (MKT).

The practice lessons. The extra step of the practice lessons has also played an important role in supporting the growth of the MKT. Considering the integer lesson alone, the addition of practice lessons resulted in discussions that not only involved analyses of learning trajectories and developmental issues for students' understanding of integers, but also the effectiveness of different representations for understanding integers and the role of contexts in learning mathematics.

This case study, while focused on a particular group in a particular situation, provides general insights and implications for both lesson study and other professional development. In a recent paper, Perry and Lewis (2009) present a case study of a successful sustained district-wide lesson study initiative. Their findings point to factors that contributed to the longevity and sustainability of that lesson study project. Amongst their findings was the crucial importance of providing teachers with diverse and frequent learning opportunities, or in Fullan's words, with *continuous learning* (Fullan 1993). In this case study, having the capacity to adapt the lesson study process *after* it was known and effective provided diverse and intellectually engaging learning opportunities for teachers and for students.

REFERENCES

Ball, Deborah Loewenberg, Mark Hoover Thames, and Geoffrey Phelps. "Content Knowledge for Teaching: What Makes It Special?" *Journal of Teacher Education* 59, no. 5 (2008): 389–407.

Bruce, Cathy, and Mary Ladky. "Using Design Research to Test and Refine a Lesson Study Model: A Close Examination of the Complexities of the Lesson Study Cycle." (Paper presented at the annual meeting of the American Educational Research Association, Denver, Colorado, April 2010.)

Collins, Allan, Diana Joseph, and Katerine Bielaczyc. "Design Research: Theoretical and Methodological Issues." *Journal of the Learning Sciences* 13, no. 1 (2004): 15–42.

Davis, Brent, and Elaine Simmt. "Mathematics-for-Teaching: An Ongoing Investigation of the Mathematics That Teachers (Need to) Know." *Educational Studies in Mathematics* 61, no. 3 (2006): 293–319.

Davis, Brent, and Elaine Simmt. "Understanding Learning Systems: Mathematics Education and Complexity Science." *Journal for Research in Mathematics Education* 34, no. 2 (2003): 137–67.

Fernandez, Clea, and Joanna Cannon. "What Japanese and U.S. Teachers Think about When Constructing Mathematics Lessons: A Preliminary Investigation." *The Elementary School Journal* 105, no. 5 (May 2005): 481–98.

Fernandez, Clea, Joanna Cannon, and Sonal Chokshi. "A US–Japan Lesson Study Collaboration Reveals Critical Lenses for Examining Practice." *Teaching and Teacher Education* 19, no. 2 (2003): 171–85.

Fernandez, Clea, and Makoto Yoshida. *Lesson Study: A Japanese Approach to Improving Mathematics Teaching and Learning.* Mahwah, N.J.: Lawrence Erlbaum Associates, 2004.

Fullan, Michael. *Change Forces: Probing the Depths of Educational Reform.* New York: Falmer Press, 1993.

Kelly, Anthony E., and Richard A. Lesh, eds. *Handbook of Research Design in Mathematics and Science Education.* Mahwah, N.J.: Lawrence Erlbaum Associates, 2000.

Lewis, Catherine. "Does Lesson Study Have a Future in the United States?" *Nagoya Journal of Education and Human Development* 1 (2002a): 1–23.

——————. *Lesson Study: A Handbook of Teacher-Led Instructional Change.* Philadelphia: Research for Better Schools, 2002b.

Lewis, Catherine C., and Ineko Tsuchida. "A Lesson Is Like a Swiftly Flowing River: How Research Lessons Improve Japanese Education." *American Educator* 22, no. 4 (Winter 1998): 12, 17, 50–52.

Lewis, Catherine, Rebecca Perry, and Aki Murata. "How Should Research Contribute to Instructional Improvement? The Case of Lesson Study." *Educational Researcher* 35, no. 3 (April 2006): 3–14.

Moss, Joan. "A Partnership of Reflective Practitioners Engages in Japanese Lesson Study for the Enhancement of Mathematics Teaching." *In School/University Partnerships: Enriching and Extending Partnerships,* edited by Carol Rolheiser, pp. 13–17. Toronto, Ont., Canada: OISE Teacher Education, 2008.

Murata, Aki. "Conceptual Overview of Lesson Study: Introduction." In *Lesson Study Research and Practice in Mathematics Education: Learning Together,* edited by Lynn C. Hart, Alice S. Alston, and Aki Murata, pp. 1–12. New York: Springer, 2011.

Perry, Rebecca R., and Catherine C. Lewis. "What Is Successful Adaptation of Lesson Study in the U.S.?" *Journal of Educational Change* 10, no. 4 (2009): 365–91.

Stigler, James W., and James Hiebert. *The Teaching Gap: Best Ideas from the World's Teachers for Improving Education in the Classroom.* New York: Free Press, 1999.

Wood, Terry, and Betsy Berry. "What Does 'Design Research' Offer Mathematics Teacher Education?" *Journal of Mathematics Teacher Education* 6, no. 3 (2003): 195–99.